WJEC
Mathematics
for AS Level – Pure

Stephen Doyle

Published in 2017 Illuminate Publishing Limited, an imprint of Hodder Education, an Hachette UK Company, Carmelite House, 50 Victoria Embankment, London EC4Y 0DZ

Orders: please contact Hachette UK Distribution, Hely Hutchinson Centre, Milton Road, Didcot, Oxfordshire, OX11 7HH. Telephone: +44 (0)1235 827827. Email: education@hachette.co.uk. Lines are open from 9 a.m. to 5 p.m., Monday to Friday. You can also order through our website: www.hoddereducation.co.uk

British Library Cataloguing in Publication Data

A catalogue record for this book is available from the British Library

ISBN 978 1 911208 51 8

Printed by Ashford Colour Press, UK

Impression 3
Year 2024

Hachette UK's policy is to use papers that are natural, renewable and recyclable products and made from wood grown in well-managed forests and other controlled sources. The logging and manufacturing processes are expected to conform to the environment regulations of the country of origin.

Editor: Geoff Tuttle
Cover design: Neil Sutton
Text design and layout: GreenGate Publishing Services, Tonbridge, Kent

Photo credits

Cover: Klavdiya Krinichnaya/Shutterstock; p9 Radachynskyi Serhii/Shutterstock; p17 Anatoli Styf/Shutterstock; p62 Jeeraphun Kulpetjira/Shutterstock; p86 Jan Miko/Shutterstock; p93 Africa Studio/Shutterstock; p115 Fotos593/Shutterstock; p135 jo Crebbin/Shutterstock; p153 Jose Antonio Perez/Shutterstock; p167 Rena Schild/Shutterstock.

Acknowledgements

The author and publisher wish to thank Rachel Bennett and Siok Barham for their help and careful attention in reviewing this book.

Contents

Contents

How to use this book

The contents of this study and revision guide are designed to guide you through to success in the Pure Mathematics component of the WJEC Mathematics for AS Level: Pure examination. It has been written by an experienced author and teacher. This book has been written specifically for the WJEC AS course you are taking and includes everything you need to know to perform well in your exams.

Knowledge and Understanding

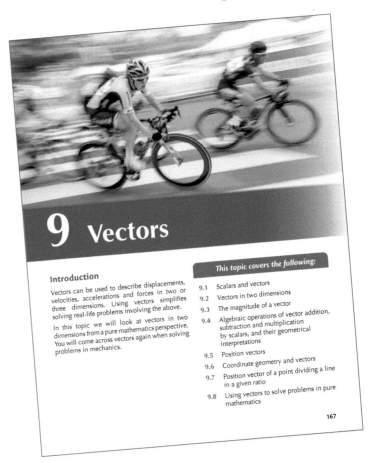

Topics start with a short list of the material covered in the topic and each topic will give the underpinning knowledge and skills you need to perform well in your exams.

The knowledge section is kept fairly short leaving plenty of space for detailed explanation of examples. Pointers will be given to the theory, examples and questions that will help you understand the thinking behind the steps. You will also be given detailed advice when it is needed.

The following features are included in the knowledge and understanding sections:

- **Grade boosts** are tips to help you achieve your best grade by avoiding certain pitfalls which can let students down.

- **Step by steps** are included to help you answer questions that do not guide you bit by bit towards the final answer (called unstructured questions). In the past, you would be guided to the final answer by the question being structured. For example, there may have been parts (a), (b), (c) and (d). Now you can get questions which ask you to go to the answer to part (d) on your own. You have to work out for yourself the steps (a), (b) and (c) you would need to take to arrive at the final answer. The 'step by steps' help teach you to look carefully at the question to analyse what steps need to be completed in order to arrive at the answer.

- **Active learning** – are short tasks which you carry out on your own which aid understanding of a topic or help with revision.

- **Summaries** – are provided for each topic and present the formulae and the main points in a topic. They can be used for quick reference or help with your revision.

Exam Practice and Technique

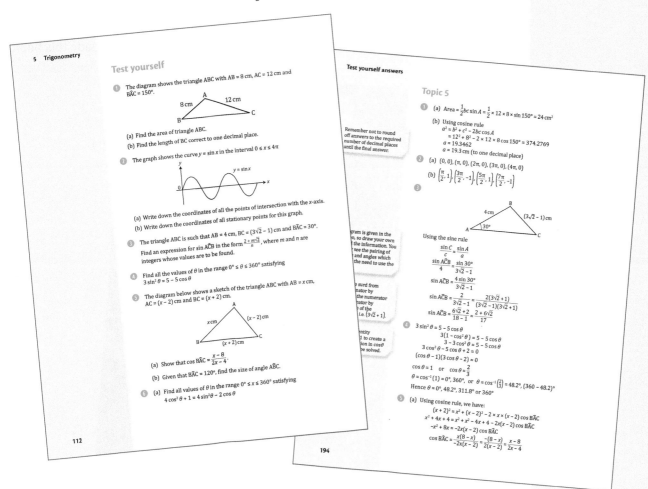

Helping you understand how to answer examination questions lies at the heart of this book. This means that we have included questions throughout the book that will build up your skills and knowledge until you are at a stage to answer full exam questions on your own. Examples are included; some of which are full examination style questions. These are annotated with Pointers and general advice about the knowledge, skills and techniques needed to answer them.

There is a Test yourself section where you are encouraged to answer questions on the topic and then compare your answers with the ones given at the back of the book. There are many examination-standard questions in each Test yourself that provide questions with commentary so you can see how the question should be answered.

You should, of course, work through complete examination papers as part of your revision process.

We advise that you look at the WJEC website www.wjec.co.uk where you can download materials such as the specification and past papers to help you with your studies. From this website you will be able to download the formula booklet that you will use in your examinations. You will also find specimen papers and mark schemes on the site.

WJEC Mathematics For AS Level Pure & Applied Practice Tests

There is another book which can be used alongside this book. This book provides extra testing on each topic and provides some exam style test papers for you to try. I would strongly recommend that you get a copy of this and use it alongside this book.

Good luck with your revision.

Stephen Doyle

1 Proof

Introduction

A mathematical proof is a mathematical argument that convinces others that a mathematical statement is true. In this topic, you will be using a number of techniques to prove or disprove mathematical statements. This topic is unusual in mathematics as it depends on precise use of English as well as mathematical techniques.

1.1 Real and imaginary numbers

Imaginary numbers are being mentioned here as a way of illustrating when a number is not real. Imaginary numbers will not be assessed in GCE Maths AS Unit 1.

A feature of real numbers is that when they are squared, positive answers are obtained. For example, $2^2 = 4$, $(3.002)^2 = 9.012004$, $(-50)^2 = 2500$.

Numbers involving square roots of negative real numbers are called imaginary numbers. For example, $\sqrt{-5}$, $\sqrt{-9}$, $\sqrt{-53}$ are all **imaginary** numbers. The main feature of an imaginary number is that when it is squared, a negative (real) number is obtained. For example, $(\sqrt{-4})^2 = -4$, $(\sqrt{-11})^2 = -11$, $(\sqrt{-67})^2 = -67$.

The important point may be summarised as follows:

If a number m is such that $m^2 < 0$ then m **cannot** be a **real** number but is instead an imaginary number. This result will be used later.

1.2 Rational and irrational numbers

Rational numbers

A rational number is a number that can be expressed as a fraction (i.e. $\frac{a}{b}$, where $b \neq 0$) with the numerator (i.e. the top number) and the denominator (i.e. the bottom number) both being whole numbers.

All whole numbers are rational. For example, 2 is rational as it can be written as the fraction, $\frac{2}{1}$.

Fractions are always rational no matter how complicated they are, so $\frac{13209}{121342}$ is rational.

Decimals with repeating units such as 0.166666666... can be expressed as fractions (i.e. $\frac{1}{6}$ in this case) and are therefore rational.

Another way of defining a rational number is to say it can be accurately placed on a number line. For example, the number $\frac{1}{3}$ (or 0.333333...) can be accurately positioned on the number line as one third of the way between the number 0 and 1.

Irrational numbers

An irrational number is a number that cannot be expressed as a fraction. When expressed as decimals, irrational numbers have endless non-repeating numbers to the right of the decimal point. Examples of irrational numbers are π (i.e. 3.141592...) and $\sqrt{2}$ (i.e. 1.414213...).

Irrational numbers cannot be positioned on the number line.

BOOST

Grade ⇧⇧⇧⇧

There are lots of specialist terms included in this topic. In exam questions these terms will be used so make sure you understand each term and can give an example.

1.3 Proof by exhaustion

Proof by exhaustion involves using all allowable values to prove the mathematical statement is either true or false. This only works when there are only a small number of possible values to try.

Examples

1 Prove that if n is an integer between and including 1 and 5 (i.e. $1 \leq n \leq 5$), then the expression $n^2 - n + 11$ is prime.

. . . .

Answer

1 As there are a limited number of values we can use with the expression, we can enter them in turn into the expression and check that the result is a prime number.

When $n = 1,$ $n^2 - n + 11 = 1^2 - 1 + 11 = 11$

            $n = 2,$ $n^2 - n + 11 = 2^2 - 2 + 11 = 13$

            $n = 3,$ $n^2 - n + 11 = 3^2 - 3 + 11 = 17$

            $n = 4,$ $n^2 - n + 11 = 4^2 - 4 + 11 = 23$

            $n = 5,$ $n^2 - n + 11 = 5^2 - 5 + 11 = 31$

All the results are prime numbers, so the statement is true.

> Note that by entering all the possible values we have exhausted all the values. Hence if all the results are prime, then the statement is true.

---

2 If p and q are even integers less than 6, prove that the sum and difference of p and q are divisible by two.

. . . .

Answer

2 Possible values for p are 2 and 4 and possible values for q are also 2 and 4.

We can construct a table to list all the possible values:

p	q	$p + q$	$p - q$	Divisible by 2?
2	2	4	0	Yes
2	4	6	−2	Yes
4	2	6	2	Yes
4	4	8	0	Yes

As all the possible values have been exhausted and all the answers are divisible by 2, the statement is correct.

1.4 Disproof by counter-example

Disproof by counter-example allows us to prove that a property is not true by providing an example where it does not hold. For example, to disprove that 'all triangles are obtuse', we give the following disproof by counter-example: the equilateral triangle having all angles equal to sixty degrees. There are an infinite number of counter-examples we could have used here, but it only takes one of them to disprove this particular statement.

N.B. A conjecture involving the word 'all' cannot be proved by showing that it is true in one case; but it can be disproved by showing that it is untrue in one case.

Remember an irrational number cannot be expressed exactly, so you cannot locate its exact position on a number line. Irrational numbers have numbers after the decimal point which do not repeat and cannot be expressed as a fraction.

Examples

1   Here is a statement: 'for all real numbers x, if x^2 is rational, then x is rational'.

By giving a counter-example, disprove this statement.

Answer

1   Suppose $x = \sqrt{2}$ (which is an irrational number)

$x^2 = \left(\sqrt{2}\right)^2 = 2$ (which is a rational number)

We have found a counter-example where x^2 is rational but the value of x is irrational.

This means that the statement is not true for all real values of x.

2   Disprove the following mathematical statement using a suitable counter-example.

The graphs of all linear functions in a single variable are perpendicular to each other.

Answer

2   Let one function be $f(x) = 2x - 1$ and the other function be $g(x) = -3x + 2$.

These functions are both in the form $y = mx + c$.

The gradient of $f(x)$ is 2 whilst the gradient of $g(x)$ is -3.

For two straight lines to be perpendicular to each other, the product of the gradients must equal -1.

Product of gradients in this case $= 2 \times (-3) = -6$.

This proves that not all linear functions are perpendicular to each other.

Hence the given mathematical statement is incorrect.

3   Show, by counter-example, that the following statement is false. 'If the integers a, b, c, d are such that a is a factor of c and b is a factor of d, then $(a + b)$ is a factor of $(c + d)$.'

Answer

3   Suppose $a = 3$ and $c = 18$ (so a is a factor of c) and suppose $b = 2$ and $d = 10$ (so that b is a factor of d).

$(a + b) = 3 + 2 = 5$ and $(c + d) = 18 + 10 = 28$

5 is not a factor of 28 so $(a + b)$ is not a factor of $(c + d)$ for this set of numbers. Hence the statement is false.

4   Disprove the following using an appropriate counter-example.

If in the quadratic equation $ax^2 + bx + c = 0$, the numbers a, b and c are real, then if b is negative, the roots of the equation will be negative.

Answer

4   If $a = 1$, $b = -2$ and $c = 1$, then the quadratic equation becomes

$$x^2 - 2x + 1 = 0$$

Factorising this equation, we obtain

$$(x - 1)(x - 1) = 0$$

Solving, gives $x = 1$, which is a positive root.

Hence the root is positive, and the counter-example proves that the statement is incorrect.

1.5 Proof by deduction

Proof by exhaustion cannot be used if there is a large number or even infinite number of values to test, so another type of proof is needed.

Proof by deduction draws a conclusion from something known or assumed. When we solve a simple equation such as $10x = 20$ we firstly assume that $10x = 20$ and that it is possible to divide both sides of an equation by a non-zero number (i.e. 10 in this case) to arrive at the deduction that $x = 2$. Proof by deduction uses algebra to decide whether a particular statement is true or false.

Examples

1   Prove that for any integer value n, if n is odd then the value of n^2 will also be odd.

Answer

1   If p is an integer then $2p$ will always be even and $2p + 1$ will always be odd.

Now we can set $n = 2p + 1$ so $n^2 = (2p + 1)^2 = 4p^2 + 4p + 1$

As $4p^2 + 4p$ will always be even (as it can be divided exactly by 2) adding one to it will make it odd. Hence $4p^2 + 4p + 1$ will be odd. So if n is odd then the value of n^2 will also be odd.

Hence the statement has been proved.

2   Use proof by deduction to prove the statement 'if a and b are even integers, the sum and difference of a and b are divisible by 2'.

Answer

2   For integers less than 6, this was proved by exhaustion earlier, but the general case can be proved by deduction in the following way:

As a and b are even integers they both have 2 as factors and can be written in the following ways:

$a = 2p$ and $b = 2q$ where p and q are integers.

Hence $a + b = 2p + 2q = 2(p + q)$ which has 2 as a factor and so must be even.

And $a - b = 2p - 2q = 2(p - q)$ which has 2 as a factor and so must be even.

So the statement has been proved.

Squaring both sides removes the square root from the right-hand side of the equation.

3   Prove the following proposition:

If a, b are positive real numbers, then $a^2 + b^2 \geq 2ab$.

Answer

3   Assuming that positive real numbers a, b exist such that $a + b \geq 2\sqrt{ab}$

$$a^2 + b^2 \geq 2ab$$

$$(a + b)^2 - 2ab \geq 2ab$$

$$(a + b)^2 \geq 4ab$$

$$a^2 + 2ab + b^2 \geq 4ab$$

$$a^2 - 2ab + b^2 \geq 0$$

$$a^2 + b^2 \geq 2ab$$

Note this will always be true if a and b are both real numbers, because when squaring a real number, you always obtain a number ≥ 0.

BOOST
Grade ⇧⇧⇧⇧

Examination questions may not always mention the method of proof to be used so you will have to decide the method to use.

● If there are a limited and small number of values to try in order to establish whether a statement or conjecture is true or false, use proof by exhaustion.

● If it looks like a proof that can be solved algebraically, use proof by deduction.

● If the proof has an infinite number of values to test, then try to find values that do not work and use disproof by counter-example.

Active Learning

If you are still having trouble getting to grips with the different proofs, take a look at some of the videos on YouTube which explain the three proofs: Proof by exhaustion, Disproof by counter-example, Proof by deduction. Try to find some which only apply to AS-level.

Step by STEP

Here is a statement: 'Every integer that is a perfect cube is a multiple of 9 or one more than a multiple of 9 or one less than a multiple of 9'.

Steps to take

1   There are three different proofs we could use so we need to select the one to use.

2   As there are three different cases to try (i.e. cube is a multiple of 9 or one more than a multiple of 9 or one less than a multiple of 9) we can use proof by exhaustion.

3   Think about how to form three equations that can be used to test the three cases.

4   Think about what letters you will use for the values that vary (e.g. n can be the integer you start with).

Answer

Each cube number is a cube of some integer n.

All integers can be described as either a multiple of 3 or one less or two less than a multiple of 3.

Hence we have 3 cases to test which will cover all the possible integers so we use proof by exhaustion

Case 1: If $n = 3p$, then $n^3 = 27p^3$, which is a multiple of 9.

Case 2: If $n = 3p - 1$, then $n^3 = 27p^3 - 27p^2 + 9p - 1$, which is 1 less than a multiple of 9.

Case 3: If $n = 3p - 2$, then $n^3 = 27p^3 - 54p^2 + 36p - 8$
$= 27p^3 - 54p^2 + 36p - 9 + 1$, which is 1 more than a multiple of 9.

So the statement has been proved true for all integers.

Test yourself

1. Prove that if n is an integer where, $2 \leq n \leq 7$ then $n^2 + 2$ is not divisible by 4.

2. Show, by counter-example, that the statement 'If $|a + 1| = |b + 1|$, then $a = b$' is false.

3. Using proof by deduction, prove that for any four consecutive integers, the difference in the products of the last two numbers and the product of the first two numbers is equal to the sum of all four integers.

4. Using disproof by counter-example prove that the proposition 'if a, b are real numbers, then $a + b \geq 2\sqrt{ab}$' is false.

5. Using proof by deduction prove that the statement 'every perfect square is either a multiple of 3 or one more than a multiple of 3' is true.

6. Disprove each of the following statements by counter-example.
   (a) For all real numbers a and b, if $b^2 > a^2$, then $b > a$.
   (b) For all real numbers x, y and z, if $x > y$, then $xz > yz$.

7. Disprove the following statement by giving a counter-example.

   If n is an integer and n^2 is divisible by 4, then n is divisible by 4.

> The two vertical lines show that we take the modulus of the value. The modulus of a value will always change a negative value into a positive value. For example $|-3 + 1| = |-2| = 2$.

Summary

Check you know the following facts:

Real numbers are those numbers that give a positive number when squared.

Imaginary numbers when squared give a negative number (e.g. $\sqrt{-1}$ when squared gives -1).

Rational numbers can be expressed as fractions.

Irrational numbers cannot be expressed as a fraction (e.g. π, $\sqrt{2}$, $\sqrt{3}$).

To prove or disprove statements:

- Use proof by exhaustion when it is possible to put all the values into a function/ expression to prove whether a statement is true or false.

- Use proof by deduction if it is possible to use algebra to decide whether a particular statement is true or false.

- Use disproof by counter-example if you can find an example that is an exception to the statement or general rule.

2 Algebra and functions

Introduction

This topic builds on and reinforces your GCSE work on algebra and functions. It will also cover new topics. Algebra and functions are an essential part of mathematics and a good grounding in this material will be essential to progress in the subject. There is a lot of content in this topic and it will be assumed that you have a good grasp of the material from your GCSE work but have a look back at your notes or any GCSE book to familiarise yourself before you start this topic.

This topic covers the following:

2.1 Laws of indices

2.2 Use and manipulation of surds

2.3 Completing the square

2.4 Solution of quadratic equations

2.5 The discriminant of a quadratic function

2.6 Sketching the graph of a quadratic function

2.7 Simultaneous equations

2.8 Solving linear and quadratic inequalities

2.9 Using set notation for solutions of inequalities

2.10 Algebraic manipulation of polynomials (expanding brackets, factorisation, algebraic division, the remainder theorem, the factor theorem, factorising a polynomial)

2.11 Sketching curves of functions

2.12 Interpreting algebraic solutions of equations graphically

2.13 Using intersection points of graphs of curves to solve equations

2.14 Proportional relationships and their graphs

2.15 Transformations of the graph of $y = f(x)$

2.1 Laws of indices

Indices are powers and you need to be able to use them in other areas such as differentiation and integration.

Indices are easy – all you have to do is follow some rules. All the rules apply to numbers of the same base. For example, the rules would apply to $2^5 \times 2^4$ because the bases are the same (i.e. 2). They would not apply to $2^3 \times 5^4$ where the bases are different (i.e. 2 and 5).

Multiplying with indices

You simply add the indices like this:

$$2^3 \times 2^5 = (2 \times 2 \times 2) \times (2 \times 2 \times 2 \times 2 \times 2) = 2^{3+5} = 2^8$$

Remember that: $2 = 2^1$, so

$$2 \times 2^5 \times 2^{-3} = 2^{1+5+(-3)} = 2^3$$

Generally

$$a^m \times a^n = a^{m+n}$$

Dividing with indices

You simply subtract the indices. Make sure you subtract the bottom index from the top index like this:

$$3^7 \div 3^2 = \frac{3 \times 3 \times 3 \times 3 \times 3 \times 3 \times 3}{3 \times 3} = 3^{7-2} = 3^5$$

$$\frac{2^4}{2^3} = 2^{(4-3)} = 2^1 = 2$$

$$\frac{5^5}{5^7} = 5^{(5-7)} = 5^{-2}$$

$$a^m \div a^n = a^{m-n}$$

In the third example, many students make the mistake of subtracting 5 from 7 to give positive 2. Remember that the final index does not always have to be positive.

Power raised to a power

You simply multiply the indices inside and outside the bracket like this:

$$(2^3)^5 = 2^{3 \times 5} = 2^{15}$$
$$\left(2^{\frac{1}{2}}\right)^4 = 2^{\left(\frac{1}{2} \times 4\right)} = 2^2$$
$$(2^{-2})^3 = 2^{(-2 \times 3)} = 2^{-6}$$

Generally

$$(a^m)^n = a^{m \times n} = a^{mn}$$

Negative and fractional powers

A negative power means one divided by the number or letter raised to the positive power like this:

$$2^{-5} = \frac{1}{2^5}$$

$$x^{-2} = \frac{1}{x^2}$$

Generally

$$a^{-m} = \frac{1}{a^m} \quad \text{provided } (a \neq 0)$$

Fractional powers mean roots. If the denominator (i.e. bottom number) is a 2 then it is a square root and if it is a 3 then it is a cube root. For example:

$$4^{\frac{1}{2}} = \sqrt{4} = 2$$
$$8^{\frac{1}{3}} = \sqrt[3]{8} = 2$$

Where there is a fraction with a numerator (i.e. top number) larger than 1, the number inside the root is raised to the power of the numerator.

In the following example x is raised to the power $\frac{2}{3}$. The denominator (i.e. 3) means the cube root of x and the numerator (i.e. 2) means that the x is squared inside the root. It does not matter whether the cube rooting or the squaring is done first.

$$x^{\frac{2}{3}} = \sqrt[3]{x^2} = \left(\sqrt[3]{x}\right)^2$$

$$8^{\frac{2}{3}} = \sqrt[3]{8^2} = 2^2 = 4$$

or

$$8^{\frac{2}{3}} = \left(\sqrt[3]{8}\right)^2 = 2^2 = 4$$ It is easier to find the cube root first and then square the answer as shown here.

Generally

$$a^{\frac{m}{n}} = \sqrt[n]{a^m} = \left(\sqrt[n]{a}\right)^m$$

If you have a number raised to a fractional power, you need to change it to roots and powers as shown here. It does not matter whether you perform the root first and then raise the answer to the power, or the other way around. Do whichever is easiest. As a general rule, finding the root first will keep the numbers low and therefore more recognisable.

Combinations of fractional and negative powers

A negative power means the reciprocal (i.e. 1 over the number raised to the positive power) and the fraction means a root.

$$27^{-\frac{1}{3}} = \frac{1}{27^{\frac{1}{3}}} = \frac{1}{\sqrt[3]{27}} = \frac{1}{3}$$

$$x^{-\frac{3}{2}} = \frac{1}{\sqrt{x^3}}$$

$$16^{-\frac{3}{2}} = \frac{1}{\left(\sqrt{16}\right)^3} = \frac{1}{4^3} = \frac{1}{64}$$

Generally

$$a^{-\frac{m}{n}} = \frac{1}{a^{\frac{m}{n}}} = \frac{1}{\sqrt[n]{a^m}} \quad \text{or} \quad \frac{1}{\left(\sqrt[n]{a}\right)^m}$$

Here you have to find $\left(\sqrt{16}\right)^3$. Because 16 is a perfect square, it is easier to find the square root of 16 and then cube the answer rather than cube 16 and then have to square root the answer.

Zero powers

Any number raised to a zero power is always 1, even if you do not know what the number is. For example

$$x^0 = 1 \quad \text{or} \quad (ab)^0 = 1.$$
$$3^0 = 1$$
$$0.5^0 = 1$$

Generally

$$\text{If } a \neq 0, \quad a^0 = 1$$

Examples

1   Write the following equation using indices:

$$y = \frac{3}{4}\sqrt[3]{x^2} - \frac{6}{x^2} + 1$$

. . . .

Answer

1   $y = \frac{3}{4}x^{\frac{2}{3}} - 6x^{-2} + 1$

---

2   Given that $y = 8x^{-2} + \frac{3}{2}x^{\frac{1}{2}}$, find y when $x = 4$

. . . .

Answer

2   Notice that if you are substituting numbers into the above equation, you need to change from index form into roots, etc. This makes it easier to substitute the numbers in.

$$y = \frac{8}{x^2} + \frac{3}{2\sqrt{x}} = \frac{8}{4^2} + \frac{3}{2\sqrt{4}} = \frac{1}{2} + \frac{3}{4} = 1\frac{1}{4}$$

---

3   Find the value of $\left(\frac{8}{27}\right)^{-\frac{2}{3}}$, writing your answer as a mixed number.

. . . .

Answer

3   $\left(\frac{8}{27}\right)^{-\frac{2}{3}} = \dfrac{1}{\left(\frac{8}{27}\right)^{\frac{2}{3}}} = \dfrac{1}{\sqrt[3]{\left(\frac{8}{27}\right)^2}} = \dfrac{1}{\left(\frac{2}{3}\right)^2} = \dfrac{1}{\left(\frac{4}{9}\right)} = \frac{9}{4} = 2\frac{1}{4}$

2.2   Use and manipulation of surds

Numbers like $\sqrt{18}$ are called surds. Surds are irrational numbers. This means that they cannot be expressed as fractions, recurring decimals or terminating decimals. Surds can be simplified like this:

$$\sqrt{18} = \sqrt{9 \times 2} = 3\sqrt{2}$$

Here the number 18 is written as the product of two factors that include a square number. 9 is a perfect square so can be square-rooted to give a whole number answer. So $\sqrt{18} = 3\sqrt{2}$.

Always try to find the largest square factor. For example $\sqrt{80}$ could be written as $\sqrt{16 \times 5} = 4\sqrt{5}$ or $\sqrt{4 \times 20}$ but this still needs further simplification to $\sqrt{4 \times 4 \times 5} = 4\sqrt{5}$. It is quicker to spot that 16 is the highest square factor of 80, so we have:

$$\sqrt{80} = \sqrt{16 \times 5} = 4\sqrt{5}.$$

Simplifying surds

Here are some general rules when manipulating surds:

$$\sqrt{a} \times \sqrt{a} = a$$

$$\sqrt{a} \times \sqrt{b} = \sqrt{ab}$$

$$\left(\sqrt{a} + \sqrt{b}\right)\left(\sqrt{a} - \sqrt{b}\right) = a - b$$

The following examples show ways in which surds can be simplified:

1 $\left(\sqrt{3}\right)^2 = \sqrt{3} \times \sqrt{3} = 3$

2 $\left(5\sqrt{2}\right)^2 = 5\sqrt{2} \times 5\sqrt{2} = 25 \times 2 = 50$

3 $\left(3\sqrt{2}\right) \times \left(4\sqrt{2}\right) = 12 \times 2 = 24$

4 $3\sqrt{2} + 2\sqrt{2} = 5\sqrt{2}$

5 $\left(2 + \sqrt{7}\right)\left(2 + \sqrt{7}\right) = 2\left(2 + \sqrt{7}\right) + \sqrt{7}\left(2 + \sqrt{7}\right) = 4 + 2\sqrt{7} + 2\sqrt{7} + 7 = 11 + 4\sqrt{7}$

6 $\left(1 + \sqrt{3}\right)\left(5 - \sqrt{12}\right) = 1\left(5 - \sqrt{12}\right) + \sqrt{3}\left(5 - \sqrt{12}\right)$

$$= 5 - \sqrt{12} + 5\sqrt{3} - \sqrt{3 \times 12}$$

$$= 5 - 2\sqrt{3} + 5\sqrt{3} - \sqrt{36}$$

$$= -1 + 3\sqrt{3}$$

Rationalising surds

If you have a fraction with a surd on the bottom, then it needs to be removed (i.e. rationalised). This is done by multiplying the top (i.e. numerator) and bottom (i.e. denominator) of the fraction by the surd. Rationalising makes sure that the denominator is no longer an irrational number.

$$\frac{1}{\sqrt{3}} = \frac{1}{\sqrt{3}} \times \frac{\sqrt{3}}{\sqrt{3}} = \frac{\sqrt{3}}{3}$$

The fraction is simplified when there are no surds in the denominator.

When there is a fraction containing a denominator like this $\frac{1}{1-\sqrt{2}}$ to remove the irrational number in the denominator, both the numerator (i.e. top) and denominator (i.e. bottom) of the fraction are multiplied by the conjugate of the denominator which in this case is $1 + \sqrt{2}$. The conjugate is the same as the denominator except the sign is the opposite.

Hence $\dfrac{1}{1 - \sqrt{2}} = \dfrac{1}{1 - \sqrt{2}} \times \dfrac{\left(1 + \sqrt{2}\right)}{\left(1 + \sqrt{2}\right)} = \dfrac{1 + \sqrt{2}}{1 - 2} = \dfrac{1 + \sqrt{2}}{-1} = -1 - \sqrt{2}$

> Suppose you had $a + \sqrt{b}$. To find the conjugate of this expression you reverse the sign between the a and $\sqrt{b}$ so the conjugate is $a - \sqrt{b}$. If you had $-2 + 3\sqrt{5}$, the conjugate would be $-2 - 3\sqrt{5}$.

> Remember that square rooting and squaring are opposite processes.

Examples

In these questions you have been asked to simplify. In each case this is done by rationalising the denominator (i.e. by removing the surds from the denominator) and simplifying the result.

1 Simplify $\dfrac{10}{\sqrt{5}}$

. . . .

Answer

1 $\dfrac{10}{\sqrt{5}} = \dfrac{10}{\sqrt{5}} \times \dfrac{\sqrt{5}}{\sqrt{5}} = \dfrac{10\sqrt{5}}{5} = 2\sqrt{5}$

2   Simplify $\dfrac{1}{2 - \sqrt{5}}$

Rationalise the denominator by multiplying the numerator and denominator by the conjugate of the denominator.

Answer

2    $\dfrac{1}{(2 - \sqrt{5})} \dfrac{(2 + \sqrt{5})}{(2 + \sqrt{5})} = \dfrac{(2 + \sqrt{5})}{4 - 5} = \dfrac{2 + \sqrt{5}}{-1} = -2 - \sqrt{5}$

3   Simplify $\sqrt{45} + \sqrt{80} + \sqrt{125}$

Remember to spot those factors that are perfect squares.

Answer

3    $\sqrt{45} + \sqrt{80} + \sqrt{125} = \sqrt{9 \times 5} + \sqrt{16 \times 5} + \sqrt{25 \times 5} = 3\sqrt{5} + 4\sqrt{5} + 5\sqrt{5} = 12\sqrt{5}$

4   Rationalise $\dfrac{a}{\sqrt{a} + \sqrt{b}}$

Answer

4    $\dfrac{a}{\sqrt{a} + \sqrt{b}} = \dfrac{a}{\sqrt{a} + \sqrt{b}} \times \dfrac{\sqrt{a} - \sqrt{b}}{\sqrt{a} - \sqrt{b}} = \dfrac{a\sqrt{a} - a\sqrt{b}}{a - b} = \dfrac{a(\sqrt{a} - \sqrt{b})}{a - b}$

5   Simplify $\dfrac{3}{\sqrt{3}} + \sqrt{75} + (\sqrt{2} \times \sqrt{6})$

Answer

5    $\dfrac{3}{\sqrt{3}} + \sqrt{75} + (\sqrt{2} \times \sqrt{6}) = \dfrac{3 \times \sqrt{3}}{\sqrt{3} \times \sqrt{3}} + \sqrt{25 \times 3} + (\sqrt{2} \times \sqrt{2 \times 3}) = \sqrt{3} + 5\sqrt{3} + 2\sqrt{3} = 8\sqrt{3}$

6   Simplify

(a) $\dfrac{5\sqrt{7} - \sqrt{3}}{\sqrt{7} - \sqrt{3}}$

(b) $(\sqrt{15} \times \sqrt{20}) - \sqrt{75} - \dfrac{\sqrt{60}}{\sqrt{5}}$

Answer

Remember that $\sqrt{7}\sqrt{3} = \sqrt{3}\sqrt{7} = \sqrt{21}$ so the middle two terms can be subtracted.

6   (a) $\dfrac{5\sqrt{7} - \sqrt{3}}{\sqrt{7} - \sqrt{3}} = \dfrac{(5\sqrt{7} - \sqrt{3})(\sqrt{7} + \sqrt{3})}{(\sqrt{7} - \sqrt{3})(\sqrt{7} + \sqrt{3})}$

Remember you have to multiply top and bottom by the conjugate of the bottom.

$= \dfrac{5 \times 7 + 5\sqrt{7}\sqrt{3} - \sqrt{3}\sqrt{7} - 3}{7 + \sqrt{7}\sqrt{3} - \sqrt{3}\sqrt{7} - 3}$

Notice that the 4 on the bottom can be divided into both terms on the top.

$= \dfrac{35 + 4\sqrt{7}\sqrt{3} - 3}{4} = \dfrac{32 + 4\sqrt{7}\sqrt{3}}{4}$

$= 8 + \sqrt{21}$

$\sqrt{5}$ is cancelled in the fraction.

(b) $(\sqrt{15} \times \sqrt{20}) - \sqrt{75} - \dfrac{\sqrt{60}}{\sqrt{5}} = \sqrt{300} - \sqrt{3 \times 25} - \dfrac{\sqrt{5 \times 12}}{\sqrt{5}}$

60 is split into the factors 5 and 12. These are chosen as there is a $\sqrt{5}$ on the bottom which will cancel.

The numbers inside the roots are written as the product of two factors where one of the factors is a perfect square.

$= \sqrt{3 \times 100} - \sqrt{3 \times 25} - \sqrt{12}$

$= 10\sqrt{3} - 5\sqrt{3} - \sqrt{4 \times 3}$

$= 10\sqrt{3} - 5\sqrt{3} - 2\sqrt{3}$

$= 3\sqrt{3}$

7 Show that $\left(\sqrt{3} - \sqrt{2}\right)^3$ can be expressed as $a\sqrt{3} - b\sqrt{2}$ where a and b are integers.

Answer

Expanding the last two brackets first.

7
$$\left(\sqrt{3} - \sqrt{2}\right)^3 = \left(\sqrt{3} - \sqrt{2}\right)\left(\sqrt{3} - \sqrt{2}\right)\left(\sqrt{3} - \sqrt{2}\right)$$
$$= \left(\sqrt{3} - \sqrt{2}\right)\left(3 - \sqrt{6} - \sqrt{6} + 2\right)$$
$$= \left(\sqrt{3} - \sqrt{2}\right)\left(5 - 2\sqrt{6}\right)$$
$$= 5\sqrt{3} - 2\sqrt{18} - 5\sqrt{2} + 2\sqrt{12}$$
$$= 5\sqrt{3} - 2\sqrt{9 \times 2} - 5\sqrt{2} + 2\sqrt{4 \times 3}$$
$$= 5\sqrt{3} - 6\sqrt{2} - 5\sqrt{2} + 4\sqrt{3}$$
$$= 9\sqrt{3} - 11\sqrt{2}$$

Look for factors of 18 and 12 which are perfect squares.

8 Simplify

(a) $\dfrac{4\sqrt{2} - \sqrt{11}}{3\sqrt{2} + \sqrt{11}}$

(b) $\dfrac{7}{2\sqrt{14}} + \left(\dfrac{\sqrt{14}}{2}\right)^3$

Answer

8 (a)
$$\frac{4\sqrt{2} - \sqrt{11}}{3\sqrt{2} + \sqrt{11}} = \frac{4\sqrt{2} - \sqrt{11}}{3\sqrt{2} + \sqrt{11}} \times \frac{3\sqrt{2} - \sqrt{11}}{3\sqrt{2} - \sqrt{11}}$$
$$= \frac{24 - 4\sqrt{22} - 3\sqrt{22} + 11}{18 - 11} = \frac{35 - 7\sqrt{22}}{7} = 5 - \sqrt{22}$$

(b)
$$= \frac{7}{2\sqrt{14}} + \left(\frac{\sqrt{14}}{2}\right)^3 = \frac{7}{2\sqrt{14}} \times \frac{\sqrt{14}}{\sqrt{14}} + \frac{14\sqrt{14}}{8} = \frac{\sqrt{14}}{4} + \frac{7\sqrt{14}}{4} = 2\sqrt{14}$$

2.3 Completing the square

A quadratic expression $x^2 + bx + c$ can be written in the form $(x + p)^2 + q$ and this is called completing the square. Note that the values of p and q can be positive or negative.

Completing the square is a useful technique for solving quadratic equations, finding the maximum or minimum value of an expression or finding the turning points of a curve.

For example, suppose you were asked to express $x^2 + 6x + 11$ in the form $(x + a)^2 + b$, where the values of a and b are to be determined.

Provided there is no number other than 1 in front of the x^2 (called the coefficient of x^2), halve the number in front of the x (the coefficient of x). Here there is a 6 in front of the x so halving this gives 3. If there is a minus sign, then this will need to be included.

The number including the sign is substituted into a bracket like this:

$$(x + 3)^2$$

When this is expanded it gives $x^2 + 6x + 9$. So we have the first two terms and also a number 9 which we remove by subtracting outside the bracket like this:

$$(x + 3)^2 - 9$$

It is now necessary to add the 11, so we have

$$x^2 + 6x + 11 = (x + 3)^2 - 9 + 11 = (x + 3)^2 + 2$$

The answer can be compared with $(x + a)^2 + b$

Hence $a = 3$ and $b = 2$

When the coefficient of x^2 is not equal to 1, we use the following method:

Example

1 Express $2x^2 + 12x + 3$ in the form $a(x + b)^2 + c$, where a, b and c are to be determined.

Answer

Before completing the square, take 2 out as a factor because the coefficient of x^2 needs to be one when completing the square.

1 $2x^2 + 12x + 3 = 2\left[x^2 + 6x + \dfrac{3}{2}\right]$

$= 2\left[(x + 3)^2 - 9 + \dfrac{3}{2}\right]$ We now complete the square inside the square bracket.

$= 2\left[(x + 3)^2 - \dfrac{15}{2}\right]$

Now multiply by the two outside the square bracket to give the required format.

$= 2(x + 3)^2 - 15$

Hence $a = 2$, $b = 3$, $c = -15$

2.4 Solution of quadratic equations

Quadratic equations are equations that can be written in the form: $ax^2 + bx + c = 0$

There are three ways to solve quadratic equations:

1 By factorising. You should be familiar with this from your GCSE work.

2 By completing the square.

3 By using the quadratic formula.

Solving a quadratic equation by factorising

Example

1 Solve $2x^2 + 7x - 4 = 0$

Answer

1 $(2x - 1)(x + 4) = 0$

Substituting each bracket equal to 0 gives $2x - 1 = 0$ or $x + 4 = 0$

Hence $x = \dfrac{1}{2}$ or $x = -4$

Grade

Make sure you can factorise quadratic expressions and then solve quadratic equations as you will need this knowledge in many parts of the AS and A2 course.

Solving a quadratic equation by completing the square

Example

1 Show that $x^2 + 0.8x - 3.84$ may be expressed in the form $(x + p)^2 - 4$, where p is a constant whose value is to be found. Hence solve the quadratic equation $x^2 + 0.8x - 3.84 = 0$.

Answer

1 $x^2 + 0.8x - 3.84 = (x + 0.4)^2 - 0.16 - 3.84$

$$= (x + 0.4)^2 - 4$$

Hence $p = 0.4$

$x^2 + 0.8x - 3.84 = 0$

So, $(x + 0.4)^2 - 4 = 0$

$(x + 0.4)^2 = 4$

Square-rooting both sides gives:

$(x + 0.4) = \pm 2$

So, $x = 2 - 0.4$ or $x = -2 - 0.4$

Hence $x = 1.6$ or $x = -2.4$

> **BOOST**
>
> **Grade** ⇧⇧⇧⇧
>
> You must use 'completing the square' to solve the quadratic equation. You will lose marks if a method is specified in the question and you use a different method.

> You must include both the positive and negative values when you square-root a number.

Solving a quadratic equation using the formula

Quadratic equations, when in the form $ax^2 + bx + c = 0$ can be solved using the formula:

$$x = \frac{-b \pm \sqrt{b^2 - 4ac}}{2a}$$

> Be careful with signs when you are entering numbers into this equation.

Example

1 Solve the equation $2x^2 - x - 6 = 0$

Answer

1 Comparing the equation given with $ax^2 + bx + c = 0$ gives $a = 2, b = -1, c = -6$. Substituting these values into the quadratic equation formula gives:

$$x = \frac{1 \pm \sqrt{(-1)^2 - 4(2)(-6)}}{2(2)}$$

$$= \frac{1 \pm \sqrt{1 + 48}}{4} = \frac{1 \pm 7}{4} = \frac{1 + 7}{4} \text{ or } \frac{1 - 7}{4} = 2 \text{ or } -1.5$$

> **Important Note:** This formula will not be given in the formula booklet so you will need to remember it.

> If you get asked to leave your answer in surd form, then you must use either the formula or the method involving completing the square.

2.5 The discriminant of a quadratic function

The roots of a quadratic equation are the same as the solutions and are also the x-coordinates of the points where the graph of the equation cuts the x-axis.

$ax^2 + bx + c =$ is a quadratic function. The quantity $b^2 - 4ac$ is called the discriminant and it gives the following information about the roots of the quadratic equation $ax^2 + bx + c = 0$:

If $b^2 - 4ac > 0$, then there are two real and distinct (i.e. different) roots.

If $b^2 - 4ac = 0$, then there are two real and equal roots.

If $b^2 - 4ac < 0$, then there are no real roots.

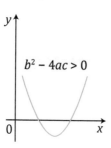 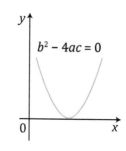 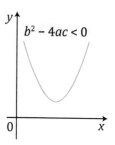

2.6 Sketching the graph of a quadratic function

The quadratic equation having the form $y = ax^2 + bx + c$ has a graph which is a parabola. Depending on the sign of a in the above equation, the parabola is ∪-shaped if a is positive or ∩-shaped if a is negative.

To find the points where the parabola intersects the x-axis, you can solve the equation $ax^2 + bx + c = 0$.

If a question starts by asking you to complete the square and then later in the question asks you to sketch the curve and/or find the maximum or minimum value then there is a quick way of doing this.

When the square has been completed, the equation for the curve will be in this format:

$$y = a(x + p)^2 + q$$

When $x = -p$, the value of the bracket is zero and, since the bracket is squared, this is its minimum value (since it cannot be negative), hence the minimum value of y is q.

If a is positive (i.e. $a > 0$) the curve will be ∪-shaped.

If a is negative (i.e. $a < 0$) the curve will be ∩-shaped.

The vertex (i.e. the maximum or minimum point) will be at $(-p, q)$.

The axis of symmetry will be $x = -p$.

For example the curve with the equation $y = 2(x + 3)^2 - 1$ can be compared with $y = a(x + p)^2 + q$. This gives $a = 2$, $p = 3$ and $q = -1$.

The curve will be ∪-shaped with a minimum point at $(-3, -1)$, i.e. $(-p, q)$, and an axis of symmetry of $x = -3$.

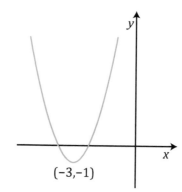

Example

1 Express $3x^2 - 12x + 17$ in the form $a(x + b)^2 + c$, where the values of the constants a, b and c are to be found.

Hence, sketch the graph of $y = 3x^2 - 12x + 17$, indicating the coordinates of its maximum or minimum point.   [5]

· ·

Answer

1
$$3x^2 - 12x + 17 = 3\left[x^2 - 4x + \frac{17}{3}\right]$$

$$= 3\left[(x - 2)^2 - 4 + \frac{17}{3}\right]$$

$$= 3\left[(x - 2)^2 + \frac{5}{3}\right]$$

$$= 3(x - 2)^2 + 5$$

> The 3 must be taken out as a factor first before completing the square.

> Compare your answer with the format for the expression in the question which in this case is $a(x + b)^2 + c$ to find the values of a, b and c.

Hence $a = 3$, $b = -2$, $c = 5$

$$y = 3x^2 - 12x + 17$$

Using your answer from completing the square gives:

$$y = 3(x - 2)^2 + 5$$

This equation is in the format $y = a(x + p)^2 + q$ where $a = 3$ (which is positive so the graph will be $\cup$-shaped). When $x = 2$, the value of the bracket is zero, so the minimum value of y is 5. The vertex (a minimum point in this case) will be at $(-p, q)$ which gives the point $(2, 5)$.

Although you are not asked to find where the curve cuts the y-axis, finding this point is useful when drawing the graph. In other questions you may be specifically asked to find the y-coordinate of this point.

To find where the curve cuts the y-axis, substitute $x = 0$ into the equation for the curve $y = 3x^2 - 12x + 17$, to give $y = 17$.

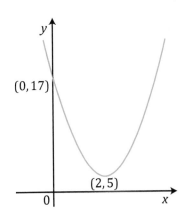

Step by STEP

1   Using the technique of completing the square, find the greatest value of

$$\frac{1}{4x^2 - 8x + 7}$$
[8]

Steps to take

1   Complete the square of the denominator: 4 will need to be taken out as a factor as we only complete the square when the number in front of the x^2 is 1.

2   Now we can find the minimum value of the denominator by finding the smallest value of $4x^2 - 8x + 7$.

3   The greatest value of $\dfrac{1}{4x^2 - 8x + 7}$ can be found when the denominator has its smallest value, so simply insert this smallest value into the above.

· ·

Answer

In order to complete the square 4 is taken out as a factor so there is a single x^2 term.

$$4x^2 - 8x + 7 = 4\left(x^2 - 2x + \frac{7}{4}\right)$$

Completing the square of the bracket gives:

$$4\left[(x - 1)^2 - 1 + \frac{7}{4}\right] = 4\left[(x - 1)^2 + \frac{3}{4}\right] = 4(x - 1)^2 + 3$$

The coordinates of the stationary point are $(1, 3)$

$$\frac{1}{4x^2 - 8x + 7} = \frac{1}{4(x - 1)^2 + 3}$$

This expression has its greatest value when the denominator has its smallest value.

The minimum value of $y = 4x^2 - 8x + 7$ is 3

The greatest value of $\dfrac{1}{4x^2 - 8x + 7}$ is when the denominator has its minimum value.

Hence the greatest value is $\frac{1}{3}$.

Example

1   (a)   Express $x^2 + 6x - 4$ in the form $(x + a)^2 + b$, where the values of a and b are to be determined.
[2]

   (b)   Use your results to part (a) to find the least value of $2x^2 + 12x - 8$ and the corresponding value of x.
[2]

· ·

Answer

1   (a) $x^2 + 6x - 4 = (x + 3)^2 - 9 - 4$

$$= (x + 3)^2 - 13$$

Hence $a = 3$ and $b = -13$

   (b) $2x^2 + 12x - 8 = 2(x^2 + 6x - 4)$

Notice that the expression in the bracket is the same as in part (a).

Using the answer to part (a) gives $2[(x + 3)^2 - 13]$.

Multiplying the contents of the square bracket by the two outside gives $2(x + 3)^2 - 26$

Now the expression is in the form $a(x + p)^2 + q$ where $a = 2$ (which is positive so the graph will be ∪-shaped). Also, the vertex (a minimum point in this case) will be at $(-p, q)$ which gives the point $(-3, -26)$.

Hence, least value is -26 and this occurs when $x = -3$.

---

2 (a) Express $x^2 - 5x + 8$ in the form $(x + a)^2 + b$, where the values of a and b are to be found.

(b) Deduce the greatest value of $-x^2 + 5x - 8$ [3]

Answer

2 (a) $x^2 - 5x + 8 = \left(x - \dfrac{5}{2}\right)^2 - \dfrac{25}{4} + 8 = \left(x - \dfrac{5}{2}\right)^2 + \dfrac{7}{4}$

Hence $a = -\dfrac{5}{2}$ and $b = \dfrac{7}{4}$

(b) $-x^2 + 5x - 8 = -(x^2 - 5x + 8) = -\left[\left(x - \dfrac{5}{2}\right)^2 + \dfrac{7}{4}\right]$

> Use the result from completing the square, and notice that all the signs are the opposite.

This function is a reflection of the function in part (a) in the x-axis (because of the minus sign).

The function in part (a) has a minimum value at $x = \dfrac{5}{2}$ of $\dfrac{7}{4}$

The reflection in the x-axis will have a maximum value of $-\dfrac{7}{4}$

2.7 Simultaneous equations

In GCSE you solved two linear equations simultaneously to find any values fitting both equations. You were finding the coordinates of the point of intersection of two straight lines.

In AS Pure you need to find the solution of one linear equation and one quadratic equation. Here you will be finding the points of intersection or the point of contact of a straight line and a curve.

Example

1 Solve the simultaneous equations $y = 10x^2 - 5x - 2$ and $y = 2x - 3$ algebraically. Write down a geometrical interpretation of your results.

Answer

1 Equating expressions for y gives

$$10x^2 - 5x - 2 = 2x - 3$$

$$10x^2 - 7x + 1 = 0$$

> At the points of intersection, the y-coordinates of the curve and straight line will be the same.

Factorising this quadratic gives

$$(5x - 1)(2x - 1) = 0$$

Hence $x = \dfrac{1}{5}$ or $x = \dfrac{1}{2}$

Substituting $x = \dfrac{1}{5}$ into $y = 2x - 3$ gives

$$y = -2\dfrac{3}{5}$$

Substituting $x = \dfrac{1}{2}$ into $y = 2x - 3$ gives

$$y = -2$$

There are two places where the line and curve intersect.

The points of intersection of the line with the curve are $\left(\dfrac{1}{5}, -2\dfrac{3}{5}\right)$ and $\left(\dfrac{1}{2}, -2\right)$.

> It is easier to substitute the x-coordinate into the equation of the straight line rather than the curve.

---

2   Solve the simultaneous equations $y = x^2 - x - 7$ and $y = 2x + 3$ algebraically. Write down a geometric interpretation of your results.   [5]

· ·

Answer

2   Equating expressions for y gives:

$$x^2 - x - 7 = 2x + 3$$

$$x^2 - 3x - 10 = 0$$

Factorising the quadratic equation gives:

$$(x - 5)(x + 2) = 0$$

Hence $x = 5$ or -2

The corresponding y-values are found by substituting these two values in $y = 2x + 3$.

Hence, when $x = 5$, $y = 13$ and when $x = -2$, $y = -1$.

The equation $y = x^2 - x - 7$ is a curve and the equation $y = 2x + 3$ is a straight line.

The points $(5, 13)$ and $(-2, -1)$ are the points where the curve and line intersect.

2.8 Solving linear and quadratic inequalities

Solving linear inequalities

These are solved in a similar way to solving ordinary linear equations but there is one important difference. If you multiply or divide both sides by a negative quantity, then the inequality sign must be reversed since, for example, $3 > 2$, but $-3 < -2$.

Examples

1 Solve the inequality $3x - 7 < 2$

. . . .

Answer

1    $3x - 7 < 2$

   $3x < 9$ (adding 7 to both sides)

   $x < 3$ (dividing both sides by 3)

---

2 Solve the inequality $1 - 2x > 5$

. . . .

Answer

2    $1 - 2x > 5$

   $-2x > 4$ (subtracting 1 from both sides)

   $x < -2$ (dividing both sides by -2 and reversing the sign)

> Whenever you multiply or divide an inequality by a negative number you must remember to reverse the inequality sign.

Solving quadratic inequalities

Examples

1 Find the range of values of x satisfying the inequality $2x^2 + x - 6 \leq 0$

. . . .

Answer

1    $2x^2 + x - 6 \leq 0$

Considering the case where $2x^2 + x - 6 = 0$ and factorising gives

   $(2x - 3)(x + 2) = 0$

Giving the critical values

   $x = \dfrac{3}{2}$ or $x = -2$ (these are the points where the curve cuts the x-axis)

Sketching the curve for $y = 2x^2 + x - 6$ gives the following:

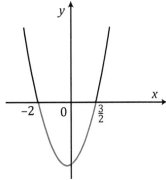

> A quick sketch of the curve and marking on it the points of intersection will enable you to see the relevant region or regions on the graph.

We want the part of the graph which is on or below the x-axis because of the $\leq$ in the inequality.

The range of values of x for which this occurs is $-2 \leq x \leq \dfrac{3}{2}$

2 Find the range of values of x satisfying the inequality $3x^2 + 2x - 1 > 0$

Answer

2   $3x^2 + 2x - 1 > 0$

Considering the case where $3x^2 + 2x - 1 = 0$ and factorising gives

$(3x - 1)(x + 1) = 0$

Giving the critical values

$x = \dfrac{1}{3}$ or $x = -1$ (these are the points where the curve cuts the x-axis)

Sketching the curve for $y = 3x^2 + 2x - 1$ gives the following:

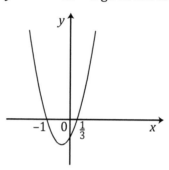

We want the part of the graph which is above the x-axis because of the $>$ in the inequality.

The range of values of x for which this occurs is $x < -1$ or $x > \dfrac{1}{3}$

> **Important note:** The value of x that satisfies the inequality is *either* less than -1 *or* greater than $\frac{1}{3}$. If you write 'and' instead of 'or', you may lose a mark.

3 (a) Given that $k \neq -1$, show that the quadratic equation

$(k + 1)x^2 + 2kx + (k - 1) = 0$

has two distinct real roots. [4]

(b) Find the range of values of x satisfying the inequality

$5x^2 + 7x - 6 \leq 0$ [3]

Answer

3   (a) Investigating the discriminant of $(k + 1)x^2 + 2kx + (k - 1) = 0$

Comparing this to $ax^2 + bx + c = 0$

gives $a = k + 1$, $b = 2k$, $c = k - 1$

Discriminant $b^2 - 4ac = (2k)^2 - 4(k + 1)(k - 1)$

$= 4k^2 - 4(k^2 - 1)$

$= 4k^2 - 4k^2 + 4$

$= 4$

> When $b^2 - 4ac > 0$ this implies that there are two real, distinct roots.

That this value is greater than 0, means there are two real and distinct roots.

(b) $5x^2 + 7x - 6 \leq 0$

Considering the case where $5x^2 + 7x - 6 = 0$
Factorising gives $(5x - 3)(x + 2) = 0$

Giving the critical values $x = \frac{3}{5}$ or -2 (these are the intercepts on the x-axis)

As the curve $y = 5x^2 + 7x - 6$ has a positive coefficient of x^2 the curve will be $\cup$-shaped.

Sketching the curve for $y = 5x^2 + 7x - 6$ gives the following:

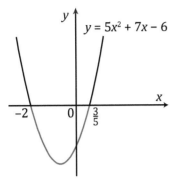

We want the part of the graph which is below or on the x-axis.

Meaning that x lies between -2 and $\frac{3}{5}$ inclusive, which can be written mathematically as $-2 \leq x \leq \frac{3}{5}$

Representing linear and quadratic inequalities graphically

Regions bounded by lines and lines, curves and lines or curves and curves can be shown graphically as a shaded area.

Strict and non-strict inequalities

A strict inequality is when an expression is > or < a value, meaning it cannot equal that value. If a straight line representing the strict inequality is drawn, it is shown as a dotted line to show that you cannot have a point on the line itself.

A non-strict inequality is when an expression is ≤ or ≥ a value meaning that it can equal the value. The line representing a non-strict inequality is drawn as a solid line showing values that lie on the line are possible.

For example, the inequality $x \geq 3$ is a non-strict inequality and can be shown graphically by drawing a solid vertical line at $x = 3$. The line is called the boundary line and points on the line itself and to the right of the line are allowable.

Examples

1 Illustrate the region represented by the following inequalities on a graph by shading the region that is required.

$$y \leq x$$

$$x + 3y \leq 12$$

$$y \geq 1$$

> This part of the topic relies on you being able to sketch lines when the equation is given. Make sure you understand how to do this.

Answer

1 First we need to add the following lines to the graph.

As all the inequalities have an equals component they are non-strict inequalities so all the lines are drawn as solid lines.

You can put any *x*-coordinate into the equation to find the *y*-coordinate.

The lines we need to draw are:

$$y = x$$
$$x + 3y = 12$$
$$y = 1$$

The line $y = 1$ is a horizontal line at $y = 1$.

For the line $y = x$, when $x = 0, y = 0$ and when $x = 4, y = 4$.

For the line $x + 3y = 12$, when $x = 0, y = 4$ and when $y = 0, x = 12$.

The lines are then added to the graph.

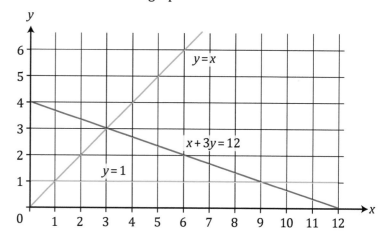

We now shade the region that is allowed.

For $y \le x$ the required region is on or below the line.

For $x + 3y \le 12$ the required region is on or below the line.

For $y \ge 1$ the required region is on or above the line.

The area where the above inequalities are allowed can be shaded as follows:

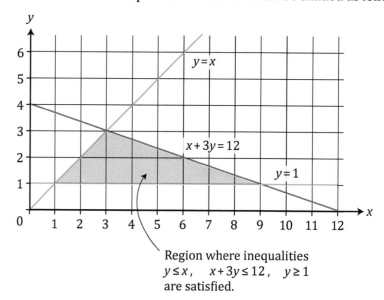

Region where inequalities
$y \le x$, $x + 3y \le 12$, $y \ge 1$
are satisfied.

It is important to note that any point which is in the feasible region or on the lines that enclose this region will satisfy all the three inequalities in the question.

2 The diagram below shows a sketch of the curve $y = 3x - x^2$ and the lines $x = 1$ and $y = -4$.

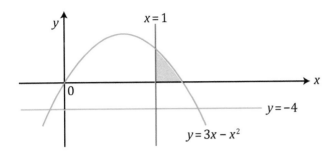

Write down three inequalities that can all be used together to describe the shaded region.

· ·

Answer

2   The shaded region is below or on the curve so $y \leq 3x - x^2$.

The shaded area is on or above the x-axis so $y \geq 0$.

The shaded area is on or to the right of the line $x = 1$, so $x \geq 1$.

Hence the area is described by $y \leq 3x - x^2$ and $y \geq 0$ and $x \geq 1$.

Notice the line $y = -4$ has no relevance and is only there as a distractor. Remember you don't have to use all the information you are given in a question.

Step by STEP

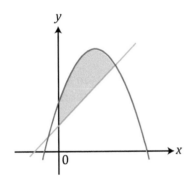

The diagram shows a sketch of the curve $y = 6 + 4x - x^2$ and the line $y = x + 2$. A point P has coordinates (a, b). Write down the three inequalities involving a and b which are such that the point P will be strictly contained within the shaded area above, if and only if, all three inequalities are satisfied. [3]

Steps to take

1   Understand the question. Initially you might think point P has been missed off from the diagram. P is not a particular point. It is any point that is in the allowable region (i.e. inside the shaded region shown on the graph).

2   The coordinates (a, b) can be substituted into the equation of the straight line and the equation of the curve to give the required inequalities.

3   Note the use of the word 'strictly' in the question. This means that none of the inequalities will contain an equals sign, meaning that values on the lines or curves are not allowed.

. . . .

Answer

If you look at the graph you can see that the area is to the right of the *y*-axis, under the curve and above the straight line. Writing the inequalities for shaded region, we have:

$a > 0$ (i.e. area is to the right of the *y*-axis)

$b < 6 + 4a - a^2$ (i.e. the area is below the curve)

$b > a + 2$ (i.e. the area is above the line)

2.9 Using set notation for solutions of inequalities

Intervals can be defined using set notation: The symbols $\{x : P\}$ designate the set of numbers *x* that satisfy condition *P*.

For example $\{x : 0 < x \le 2\}$, means the set of those numbers *x* such that $0 < x \le 2$.

This interval can also be referred to as $(0, 2]$, where the bracket at the left indicates that the point $x = 0$ is not in the interval and the square bracket at the right indicates that the point $x = 2$ is in the interval.

All those values of *x* less than or equal to 3 can be written as $\{x : x \le 3\}$ or as $(-\infty, 3]$ and all those values greater than 2 can be written as $\{x : x > 2\}$, or as $(2, \infty)$.

2.10 Algebraic manipulation of polynomials
(expanding brackets, factorisation, algebraic division, the remainder theorem, the factor theorem, factorising a polynomial)

In this section you will cover the manipulation of polynomials such as $27x^3 + 9x^2 - 3x + 7$ and also learn about the binomial expansion of $(1 + x)^n$ for positive integer values of *n*

There is a fair amount of algebraic manipulation needed, so you may need to practise some of the skills you learned as part of your GCSE course.

Expanding brackets and collecting like terms

Here are some examples of polynomials. Notice the way they are each ordered in descending powers of *x*. The degree is the highest power of *x* in the expression.

$4x - 9$ is a polynomial of degree 1 or a linear expression

$2x^2 + 4x - 1$ is a polynomial of degree 2 or a quadratic expression

$5x^3 + 3x^2 - 2x + 6$ is a polynomial of degree 3 or a cubic expression

You can only add or subtract like terms when simplifying an expression. Like terms are those terms having identical letters and powers. For example $4x^2y$ cannot be added to $5xy$.

Examples

1 $4x^3 + 6x^2y + x^2y + 5xy - xy^2 + xy - 2x^3 = 2x^3 + 7x^2y + 6xy - xy^2$

2 $a + ab + ba + b^2 - a - 4b^2 = 2ab - 3b^2$

ab and *ba* are the same and can therefore be added.

3 $x(x^2 + 2x - 1) + 2x(x - 3) = x^3 + 2x^2 - x + 2x^2 - 6x = x^3 + 4x^2 - 7x$

4   Expand and simplify the brackets in the following expression:

$(x + 2)(x - 3)(x + 4)$

$(x + 2)(x - 3)(x + 4) = (x + 2)(x^2 + 4x - 3x - 12)$

$$= (x + 2)(x^2 + x - 12)$$

$$= x^3 + x^2 - 12x + 2x^2 + 2x - 24$$

$$= x^3 + 3x^2 - 10x - 24$$

> Here we have multiplied the last pair of brackets first to give a quadratic expression. This is then multiplied by the first bracket to give the final answer. The order in which the brackets are multiplied is unimportant.

Factorisation

Factorisation is the opposite process to expanding brackets. The highest common factor is taken outside the bracket. Each term is divided by the highest common factor and the result is written inside the bracket.

Examples

Factorise the following:

1 $x^2y - xy = xy(x - 1)$

2 $24x^3y^2z + 6x^2y - 18x^2 = 6x^2(4xy^2z + y - 3)$

3 $15a^2b - 12ab = 3ab(5a - 4)$

The difference of two squares

Both terms must be perfect squares so capable of being easily square-rooted. Note that this only works when there is a minus sign between the two terms.

$$x^2 - y^2 = (x + y)(x - y)$$

$$4x^2 - 9y^2 = (2x + 3y)(2x - 3y)$$

$$16x^2 - 25 = (4x + 5)(4x - 5)$$

> Just square-root each term and substitute them in brackets like this, giving one a + sign between the terms and the other a − sign.

The following difference of two squares formula should be learned:

$$a^2 - b^2 = (a + b)(a - b)$$

Algebraic division

When 25 is divided by 4 the quotient is 6 and the remainder is 1. The number 25 can be written in the following way:

$$25 = 4 \times 6 + 1$$

This can be applied to algebra like this:

Find the quotient and remainder when $x^2 + 5x - 8$ is divided by $x - 2$.

$x^2 + 5x - 8 = (x - 2)(ax + b) + c$ where $ax + b$ is the quotient and c the remainder.

$$= ax^2 + bx - 2ax - 2b + c$$

$$= ax^2 + (b - 2a)x - 2b + c$$

BOOST

Grade ⬆⬆⬆⬆

It is always advisable to check your factorisation by multiplying out the brackets. It is easy to make a mistake especially with signs.

Comparing this with the original expression and equating the coefficients of x^2 gives $a = 1$

Equating coefficients of x gives $5 = b - 2a$ and since $a = 1$, solving gives $b = 7$.

Equating constant terms gives $-2b + c = -8$ and since $b = 7$, $c = 6$

Hence the quotient (i.e. $ax + b$) is $x + 7$ and the remainder (i.e. c) is 6.

Notice that $x^2 + 5x - 8 = (x - 2)(x + 7) + 6$

> The coefficients of x, x^2, x^3, etc. are the numbers in front of these terms. The term independent of x is the number without any x in it (i.e. the constant term).

The remainder theorem

The remainder theorem states:

> The remainder theorem will not be assessed in GCE Maths AS Unit 1 and is only included here to aid understanding.

> If a polynomial $f(x)$ is divided by $(x - a)$ the remainder is $f(a)$.

For example, if $f(x) = x^3 + 2x^2 - x + 1$ is divided by $x - 1$ the remainder will be $f(1)$.

Remainder $= f(1) = 1^3 + 2(1)^2 - 1 + 1 = 3$

Examples

1   Find the remainder when $x^3 + x^2 + x - 2$ is divided by $x - 1$

* *

Answer

1   Let $f(x) = x^3 + x^2 + x - 2$

> This is the remainder theorem.

If $f(x) = x^3 + x^2 + x - 2$ is divided by $x - 1$, the remainder is $f(1)$.

$f(1) = 1^3 + 1^2 + 1 - 2 = 1$

Hence the remainder $= 1$

---

2   Find the remainder when $27x^3 + 9x^2 - 3x + 7$ is divided by $3x - 1$.

* *

Answer

> The value of x to be substituted into the function is found by letting $3x - 1 = 0$ and then solving for x giving $x = \frac{1}{3}$

2   Let $f(x) = 27x^3 + 9x^2 - 3x + 7$

If $f(x) = 27x^3 + 9x^2 - 3x + 7$ is divided by $3x - 1$, the remainder is $f\left(\frac{1}{3}\right)$.

$f\left(\frac{1}{3}\right) = 27\left(\frac{1}{3}\right)^3 + 9\left(\frac{1}{3}\right)^2 - 3\left(\frac{1}{3}\right) + 7 = 1 + 1 - 1 + 7 = 8$

Hence the remainder $= 8$

The factor theorem

The factor theorem is in the specification.

A special case of the remainder theorem occurs when there is no remainder, i.e. when $f(a) = 0$.

> For a polynomial $f(x)$, if $f(a) = 0$ then $(x - a)$ is a factor of $f(x)$

For example in a polynomial $f(x)$, if $f(5) = 0$, then $(x - 5)$ is a factor of $f(x)$.

If for the same polynomial $f(-2) = 0$, then $(x + 2)$ is also a factor of $f(x)$.

Examples

1   Prove that $x + 3$ is a factor of the polynomial $2x^3 + x^2 - 13x + 6$

If $x + 3$ is a factor, then when $x = -3$ is substituted into $f(x)$ there will be no remainder.

Answer

1   Let $f(x) = 2x^3 + x^2 - 13x + 6$

If $x + 3$ is a factor then $f(-3)$ should be zero.

$$f(-3) = 2(-3)^3 + (-3)^2 - 13(-3) + 6 = -54 + 9 + 39 + 6 = 0$$

Hence $x + 3$ is a factor.

2   Prove that $x - 2$ is **not** a factor of the function $3x^3 - 2x^2 + x + 2$

Answer

2    $f(x) = 3x^3 - 2x^2 + x + 2$

$f(2) = 3(2)^3 - 2(2)^2 + 2 + 2 = 20$

As $f(2) \neq 0$ then $x - 2$ is not a factor of the function.

BOOST

Grade ⇧⇧⇧⇧

Always read the question carefully. It would be easy to miss the word 'not' in this question.

Factorising a polynomial

Suppose a function $f(x)$ is defined by $f(x) = x^3 - 3x^2 - x + 3$.

In order to factorise the function it is necessary first to find a factor.

Suppose we think that $(x + 1)$ is a factor. We can see if it is a factor by substituting $x = -1$ into the function. If there is no remainder then $(x + 1)$ is a factor.

$$f(-1) = (-1)^3 - 3(-1)^2 - (-1) + 3 = -1 - 3 + 1 + 3 = 0$$

Hence $(x + 1)$ is a factor.

The function can now be written in the following way:

$$f(x) = (x + 1)(ax^2 + bx + c) = x^3 - 3x^2 - x + 3$$

Equating coefficients of x^3 gives $a = 1$.

Equating constant terms gives $c = 3$.

Equating coefficients of x gives $c + b = -1$ so $b = -4$.

These values can be substituted in giving:

$$f(x) = (x + 1)(x^2 - 4x + 3)$$

The second bracket is then factorised giving:

$$f(x) = (x + 1)(x - 3)(x - 1)$$

A slightly different method of finding the quadratic factor would be to write down the x^2 and the constant term by inspection, e.g. $x^3 - 3x^2 - x + 3 = (x + 1)(x^2 + ax + 3)$ as it is clear that $x \times x^2 = x^3$ and $1 \times 3 = 3$. Then there is only one unknown coefficient to find. Equating coefficients of x^2 or x is sufficient to find this unknown coefficient of x in the quadratic factor – and doing both would act as a useful check.

Each term in the first bracket is multiplied by each term in the second bracket.

BOOST

Grade ⇧⇧⇧⇧

Three of the four terms have been equated here. The fourth term could be equated as a check. Here it is the coefficients of x^2. Equating these gives $b + a = -3$.

We can substitute the values in $b + a = -4 + 1 = -3$.

Examples

1   The polynomial $f(x)$ is defined by: $f(x) = 2x^3 + 11x^2 + 4x - 5$

  (a) (i) Evaluate $f(-2)$.

  (ii) Using your answer to part (i), write down one fact which you can deduce about $f(x)$.   [2]

  (b) Solve the equation $f(x) = 0$ [6]

. . . .

Answer

> If $f(-2) = 0$, then $(x + 2)$ is a factor. If there is a remainder, then it is not a factor.

1   (a) (i) $f(x) = 2x^3 + 11x^2 + 4x - 5$

$$f(-2) = 2(-2)^3 + 11(-2)^2 + 4(-2) - 5 = 15$$

  (ii) Since there is a remainder, this means that $(x + 2)$ is not a factor of $2x^3 + 11x^2 + 4x - 5$

  (b) Using values that are factors of the constant term, (5), we substitute values of x into the function until the function equals zero.

> Remember to reverse the sign of the number which gives a zero value, when stating the factor.

Starting from $f(1), f(-1), f(5)$ etc.

$$f(1) = 2(1)^3 + 11(1)^2 + 4(1) - 5 = 12$$

$$f(-1) = 2(-1)^3 + 11(-1)^2 + 4(-1) - 5 = 0$$

Hence $(x + 1)$ is a factor.

As one of the factors is $(x + 1)$ so the original function can be written like this:

$$2x^3 + 11x^2 + 4x - 5 = (x + 1)(ax^2 + bx + c)$$

Equating the coefficients of x^3 gives, $a = 2$.

Equating the constant terms gives, $c = -5$.

Equating the coefficients of x^2 gives,

  $b + a = 11$, so $b = 9$

Hence $2x^3 + 11x^2 + 4x - 5 = (x + 1)(2x^2 + 9x - 5)$

Factorising the quadratic part into two factors gives:

$$(x + 1)(2x - 1)(x + 5)$$

Hence $f(x) = (x + 1)(2x - 1)(x + 5) = 0$

Solutions are $x = -1, \frac{1}{2}$ or -5

BOOST

Grade ⬆⬆⬆⬆

> You must be able to factorise quadratic expressions quickly and accurately. Practise these using a GCSE text book.

> Substitute each bracket in turn equal to 0 and solve for x to obtain the solutions.

─────────────────────────────

2   (a) Find the remainder when $x^3 - 17$ is divided by $x - 3$.   [2]

  (b) Solve the equation $6x^3 - 7x^2 - 14x + 8 = 0$ [6]

. . . .

Answer

2   (a) Let $f(x) = x^3 - 17$

$$f(3) = 3^3 - 17 = 10, \text{ hence remainder} = 10$$

(b) Let $f(x) = 6x^3 - 7x^2 - 14x + 8$

$f(1) = 6(1)^3 - 7(1)^2 - 14(1) + 8 = -7$

$f(-1) = 6(-1)^3 - 7(-1)^2 - 14(-1) + 8 = 9$

$f(2) = 6(2)^3 - 7(2)^2 - 14(2) + 8 = 0$, hence $(x - 2)$ is a factor of $f(x)$

$(x - 2)(ax^2 + bx + c) = 6x^3 - 7x^2 - 14x + 8$

Equating coefficients of x^3 gives $a = 6$

Equating the constant terms gives, $-2c = 8$ so $c = -4$

Equating the coefficients of x^2 gives $b - 2a = -7$, so $b = 5$

Hence the equation is factorised to:

$(x - 2)(6x^2 + 5x - 4) = (x - 2)(3x + 4)(2x - 1)$

$(x - 2)(3x + 4)(2x - 1) = 0$

Solving gives $x = 2, -\frac{4}{3}$ or $\frac{1}{2}$.

> You have to use trial and error by substituting in values 1, −1, 2, −2, etc., until you find a value that gives zero when substituted into the function for x. Try values that are factors of the constant term (8 in this case).

3 (a) Given that $x + 2$ is a factor of $12x^3 + kx^2 - 13x - 6$, write down an equation satisfied by k. Hence show that $k = 19$. [2]

(b) Factorise $12x^3 + 19x^2 - 13x - 6$. [3]

(c) Find the remainder when $12x^3 + 19x^2 - 13x - 6$ is divided by $2x - 1$. [2]

Answer

3 (a) If $x + 2$ is a factor, then when $x = -2$ is substituted into the function, the function will equal zero.

Let $f(x) = 12x^3 + kx^2 - 13x - 6$

$f(-2) = 12(-2)^3 + k(-2)^2 - 13(-2) - 6 = 0$

$-96 + 4k + 26 - 6 = 0$

$k = 19$

(b) Notice that this is the same equation in part (a) with 19 substituted in for k. Hence we know that $(x + 2)$ is a factor.

So, $(x + 2)(ax^2 + bx + c) = 12x^3 + 19x^2 - 13x - 6$

Equating coefficients of x^3 gives $a = 12$

Equating the coefficients of x^2 gives $b + 2a = 19$

$b + 24 = 19$

$b = -5$

> With practice the values of a, b and c can be found quickly by inspection.

Equating the constant terms gives, $2c = -6$ so $c = -3$

Hence $12x^3 + 19x^2 - 13x - 6 = (x + 2)(12x^2 - 5x - 3)$

$= (x + 2)(4x - 3)(3x + 1)$

(c) $f(x) = 12x^3 + 19x^2 - 13x - 6$

$f\left(\frac{1}{2}\right) = 12\left(\frac{1}{2}\right)^3 + 19\left(\frac{1}{2}\right)^2 - 13\left(\frac{1}{2}\right) - 6 = \frac{-25}{4}$

> $2x - 1 = 0$, so $x = \frac{1}{2}$

Hence remainder $= \frac{-25}{4}$

4   (a)   Given that $x - 2$ is a factor of $x^3 - 6x^2 + ax - 6$, show that $a = 11$.   [2]

(b)   Solve the equation $x^3 - 6x^2 + 11x - 6 = 0$.   [4]

(c)   Calculate the remainder when $x^3 - 6x^2 + 11x - 6$ is divided by $x + 1$.   [2]

....

Answer

4   (a)   Let $f(x) = x^3 - 6x^2 + ax - 6$

$f(2) = 2^3 - 6(2)^2 + 2a - 6 = 2a - 22$

As $x - 2$ is a factor, $f(2) = 0$

Hence $2a - 22 = 0$, so $a = 11$

(b) $f(x) = x^3 - 6x^2 + 11x - 6 = (x - 2)(ax^2 + bx + c)$

Equating coefficients of x^3 gives $a = 1$

Equating constant terms gives, $-2c = -6$ giving $c = 3$.

Equating the coefficients of x^2 gives $b - 2a = -6$, giving $b = -4$

Hence $f(x) = x^3 - 6x^2 + 11x - 6 = (x - 2)(x^2 - 4x + 3)$

Factorising the second bracket gives:

$$f(x) = (x - 2)(x - 3)(x - 1)$$

Now $(x - 2)(x - 3)(x - 1) = 0$

So $x = 2, x = 3, x = 1$

(c) $f(-1) = (-1)^3 - 6(-1)^2 + 11(-1) - 6 = -1 - 6 - 11 - 6 = -24$

Remainder $= -24$

> Always look back at the previous part to see if it is relevant. Here it is because the polynomial is the same and we know that $x - 2$ is a factor.

5   The polynomial $4x^3 + px^2 - 11x + q$ has $x - 2$ as a factor. When the polynomial is divided by $x + 1$, the remainder is 9.

(a)   Show that $p = -4$ and $q = 6$.   [6]

(b)   Factorise $4x^3 - 4x^2 - 11x + 6 = 0$.   [3]

....

Answer

5   (a)   As $x - 2$ is a factor, when $x = 2$ is substituted into the function, the result will be zero.

Let $f(x) = 4x^3 + px^2 - 11x + q$

$f(2) = 4(2)^3 + p(2)^2 - 11(2) + q = 10 + 4p + q$

This equals zero so:

$$10 + 4p + q = 0 \tag{1}$$

Also, when the polynomial is divided by $(x + 1)$ it gives a remainder of 9

$$f(-1) = 4(-1)^3 + p(-1)^2 - 11(-1) + q$$

$$= -4 + p + 11 + q$$

$$= 7 + p + q$$

This remainder equals 9 so:

$$9 = 7 + p + q$$

$$2 = p + q \qquad (2)$$

Solving equations (1) and (2) simultaneously:

From equation (2) $q = 2 - p$

Substituting this into equation (1) gives:

$$10 + 4p + 2 - p = 0$$

$$12 + 3p = 0$$

$$p = -4$$

Substituting $p = -4$ into equation (2) gives:

$$2 = -4 + q$$

$$q = 6$$

(b)   You know that $x - 2$ is a factor of $4x^3 - 4x^2 - 11x + 6$

So $(x - 2)(ax^2 + bx + c) = 4x^3 - 4x^2 - 11x + 6$

Equating coefficients of x^3 gives $a = 4$

Equating constant terms gives, $-2c = 6$ giving $c = -3$

Equating the coefficients of x^2 gives $b - 2a = -4$

$$b - 8 = -4$$

$$b = 4$$

Substituting these values gives: $(x - 2)(4x^2 + 4x - 3)$

Factorising the second bracket gives: $(x - 2)(2x + 3)(2x - 1)$

Always look back at previous parts to see if they are relevant. Here the previous part is relevant.

The second bracket contains a quadratic which can be factorised.

2.11 Sketching curves of functions

This section looks at sketching curves defined by simple functions. You must be able to spot these simple functions and be able to immediately recognise the shape of the graph.

Graphs of $y = x^2$, $y = x^4$ and other even powers of x

These graphs are ∪-shaped and pass through the origin and look like this.

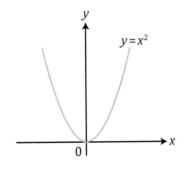

Graphs of $y = x^3$, $y = x^5$ and other odd powers of x

These graphs pass through the origin and look like this.

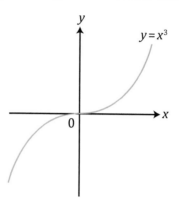

Graph of $y = \dfrac{a}{x}$ (i.e. reciprocal graphs)

The straight line which a graph approaches without actually touching is called an asymptote.

The graphs of $y = \dfrac{a}{x}$, where a is a constant all take a similar shape regardless of the value of a. These graphs all have the x and y axes as asymptotes.

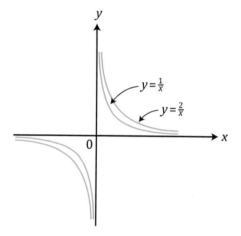

Notice as the value of a increases, the graphs have a similar shape but they lie further from the origin but still have both axes as asymptotes.

Graph of $y = \dfrac{a}{x^2}$ (i.e. reciprocal graphs)

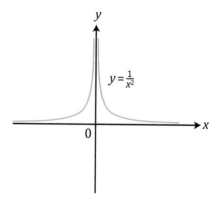

All the graphs of $y = \dfrac{a}{x^2}$ have the x-axis and the positive y-axis as asymptotes. As the value of a increases then so does the distance of the graph from the origin.

Graphs of cubic equations of the form $y = ax^3 + bx^2 + cx + d$

If the value of $a > 0$, the graphs have shapes like these going from bottom left to top right.

For some values of a, b, c and d the shape may not show a distinct peak and trough.

If the value of $a < 0$, the graphs have shapes like these going from top left to bottom right.

Sometimes the graphs do not show distinct peaks and troughs.

Again, the values of a, b, c and d determine if the graph has a peak and trough.

Some cubic functions can be factorised into the following

$$y = (x - a)(x - b)(x - c)$$

If the values of a, b and c are different, the graph of the cubic function will cut the x-axis at three places (i.e. at a, b and c).

The graph of the cubic function $y = (x + 3)(x - 1)(x - 4)$ will cut the x-axis at -3, 1 and 4. Notice the x^3 term in the expansion of the three brackets is positive so $a > 0$.

We now know the shape of the graph and where it cuts the x-axis.

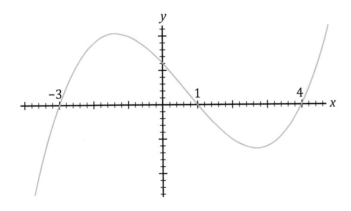

The cubic function $y = (x + 3)(x - 2)^2$ has a repeated root. When this happens the x-axis becomes a tangent to the curve at the repeated root (i.e. at $x = 2$ in this case).

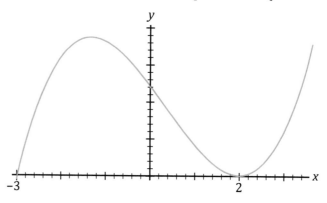

Active Learning

Did you know if you go onto Google you can type in an equation of a curve into the search box and Google will produce a graph automatically.

If you want a to produce a graph for $y = 3x^3 + 4x^2 - x - 2$ you would type it into the search box as the following $y = 3x\textasciicircum3 + 4x\textasciicircum2 - x - 2$ as shown below.

Graph for $y = 3x\textasciicircum3 + 4x\textasciicircum2 - x - 2$

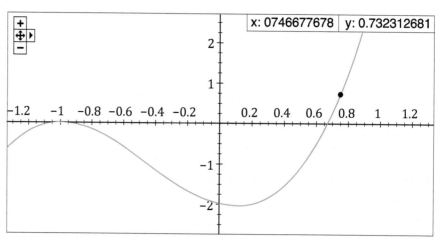

More info

Try producing the graphs of curves having the following equations:

$$y = (x - 3)^2$$
$$y = (x - 1)(x - 6)$$
$$y = (x - 1)(x + 1)(x + 4)$$
$$y = (x + 3)(x - 4)^2$$
$$y = 6x^3 + 2x^2 - x + 2$$
$$y = \frac{5}{x}$$

Step by

A cubic polynomial is given by $f(x) = 2x^3 + ax^2 + bx + c$, where a, b and c are constants.

The graph of this polynomial cuts the x-axis at -4, -2, $\frac{1}{2}$.

Find the coordinates of the point where the curve cuts the y-axis.

> A polynomial of degree 3 means the highest power of x is 3 (i.e. x^3).

Steps to take

1   This question is about polynomials and their factors. Most polynomials can be factorised into three factors.

2   As the curve cuts the x-axis in three places, there must be three factors.

3   We can multiply these factors and equate them to the given function to determine the values of a, b and c.

4   To find where the curve cuts the y-axis we can substitute $x = 0$ into the equation of the polynomial.

. . . .

Answer

Using the roots of the equation we know that the factors of the polynomial are

$$(2x - 1)(x + 4)(x + 2)$$

Hence $(2x - 1)(x + 4)(x + 2) = 2x^3 + ax^2 + bx + c$

Multiplying these brackets out we obtain

$$(2x - 1)(x^2 + 6x + 8) = 2x^3 + ax^2 + bx + c$$

$$2x^3 + 12x^2 + 16x - x^2 - 6x - 8 = 2x^3 + ax^2 + bx + c$$

$$2x^3 + 11x^2 + 10x - 8 = 2x^3 + ax^2 + bx + c$$

By comparing the coefficients, we can see that $a = 11$, $b = 10$ and $c = -8$

When $x = 0$, $f(0) = 2(0)^3 + a(0)^2 + b(0) + c = c = -8$

Hence, the curve will cut the y-axis at $(0, -8)$

> There is another method where it is not necessary to find all the coefficients. The product of the roots multiplied by coefficient of x^2 will give -8. Try it and see.
>
> This is equivalent to substituting into the factorised form of the polynomial.

2.12 Interpreting algebraic solutions of equations graphically

When two equations are solved simultaneously the set of solutions represents the coordinates of any points of intersection between the graphs of the two equations.

Example

1   (a)   Solve the equations $y = x^2 - x - 20$ and $y = -2x - 14$ simultaneously.

   (b)   State what your answers to part (a) represent.

Answer

1   (a)   Equating the y-values: $x^2 - x - 20 = -2x - 14$

$$x^2 + x - 6 = 0$$

$$(x + 3)(x - 2) = 0$$

$$x = -3 \text{ or } 2$$

When $x = -3$, $y = -2(-3) - 14 = -8$

When $x = 2$, $y = -2(2) - 14 = -18$

Solutions are $x = -3$ and $y = -8$ or $x = 2$ and $y = -18$

(b)   They are the coordinates of the points of intersection of the line and the curve.

2.13 Using intersection points of graphs of curves to solve equations

You can find the coordinates of the points where a line intersects a curve by solving the two equations simultaneously.

Example

1   Find the coordinates of the points of intersection of the line $y = 2x - 1$ with the curve $y = x^2 - 4$.

Produce on the same set of axes a sketch of the line and the curve marking on your graph the coordinates of the points of intersection of the curve with both axes. Also mark on the graph the coordinates of the points of intersection between the curve and the line.

> The y-values of both equations are equated so that the x-coordinates of the points of intersection can be found.

Answer

1   The points of intersection of the curve and line have the same y-values.

Equating the y-values gives $x^2 - 4 = 2x - 1$

$$x^2 - 2x - 3 = 0$$

$$(x - 3)(x + 1) = 0$$

Hence, $x = -1$ or 3

Substituting these values into the equation of the line we obtain

When $x = -1, y = 2(-1) - 1 = -3$

When $x = 3, y = 2(3) - 1 = 5$

Hence points of intersection are $(-1, -3)$ and $(3, 5)$

To find the points of intersection of the curve with the x-axis we let $y = 0$.

$$x^2 - 4 = 0$$

$$(x - 2)(x + 2) = 0$$

Hence, the curve cuts the x-axis at $x = 2$ and $x = -2$

To find the point of intersection of the curve with the y-axis, substitute $x = 0$ into the equation.

Hence $y = 0^2 - 4 = -4$, so the curve intersects the y-axis at $(0, -4)$

The graphs can now be sketched.

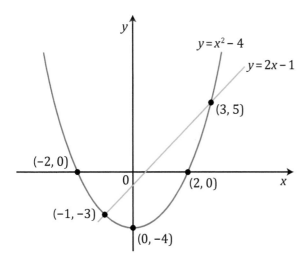

BOOST

Grade ⬆⬆⬆⬆

Check that you have included all the points of intersection asked for in the question. Make sure you give both coordinates for each point.

2.14 Proportional relationships and their graphs

Direct proportion

If y is directly proportional to x, when x doubles, triples, halves, etc., then so does y.

This can be written as $y \propto x$. The equation of the line can be written as $y = kx$ where k is a constant of proportionality which is also the gradient of the line.

The graph of y against x is a straight line through the origin.

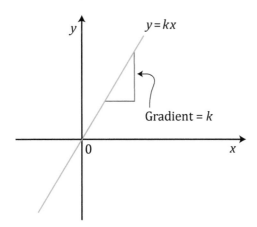

Inverse proportion

If y is inversely proportional to x this is written as $y \propto \dfrac{1}{x}$. If x doubles, y halves and so on.

This relationship written as an equation is $y = \dfrac{k}{x}$.

A graph of y against x results in the following curve.

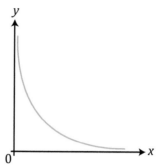

A graph of y against $\dfrac{1}{x}$ results in a straight line whose gradient is the constant of proportionality, k.

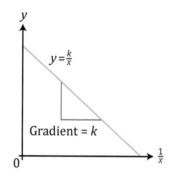

2.15 Transformations of the graph of $y = f(x)$

If you are given a graph of a function in the form $y = (x)$, then the graph of a new function may be obtained from the original graph by applying a simple transformation.

The simple transformations include reflections, translations and stretches.

$y = f(x)$ to $y = f(x + a)$

This represents a translation of $-a$ units parallel to the x-axis.

This can be represented by the translation $\begin{pmatrix} -a \\ 0 \end{pmatrix}$.

Note that if a is positive in the new function, the graph moves a units to the left and if it is negative it will move a units to the right.

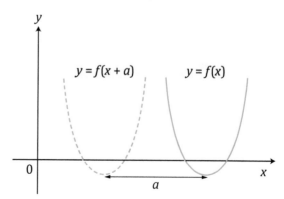

Transformations can be reflections, translations, stretches and enlargements. Rotations will not be covered here.

$y = f(x)$ to $y = f(x) + a$

This represents a translation of a units parallel to the y-axis. This can be represented by the translation $\begin{pmatrix} 0 \\ a \end{pmatrix}$.

If a is positive, the whole graph moves up a units and if a is negative it moves down by a units.

A translation simply means moving the graph so that it is exactly the same shape but in a different position.

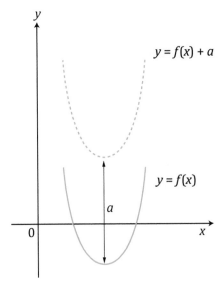

$y = f(x)$ to $y = af(x)$

This represents a one-way stretch with scale factor a parallel to the y-axis. This means that the y-value of any point on the curve will be multiplied by a leaving the x-value unchanged. It is important to note that any points of intersection with the x-axis will remain unchanged during the stretch.

Note that if a is negative the curve will be reflected in the x-axis

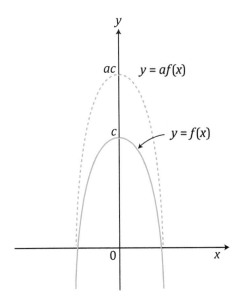

$y = f(x)$ to $y = f(ax)$

This represents a one-way stretch with scale factor $\frac{1}{a}$ parallel to the x-axis.

Notice the way each x-value on the graph for the original function is halved. Here you can see that the graph of $y = f(x)$ cuts the x-axis at $x = 2$ and $x = 4$. The graph of $y = f(2x)$ cuts the x-axis at half of these values, i.e. $x = 1$ and $x = 2$.

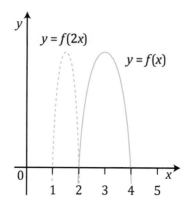

Here the x-values for the original function are doubled. The graph of $y = f(x)$ cuts the x-axis at $x = 1$ and $x = 3$. The graph of $y = f\left(\frac{1}{2}x\right)$ cuts the x-axis at double these values, i.e. $x = 2$ and $x = 6$.

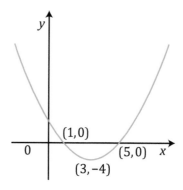

Examples

1   The diagram shows a sketch of the graph $y = f(x)$. The graph passes through the points $(1, 0)$ and $(5, 0)$ and has a minimum point at $(3, -4)$.

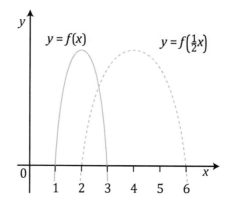

Sketch the following graphs, using a separate set of axes for each graph. In each case, you should indicate the coordinates of the stationary point and the coordinates of the points of intersection of the graph with the x-axis.

(a) $y = f(x + 1)$

(b) $y = -2f(x)$

Answer

1 (a) $y = f(x + 1)$ is a translation of $y = f(x)$ by one unit to the left parallel to the x-axis, i.e. a translation of $\begin{pmatrix} -1 \\ 0 \end{pmatrix}$

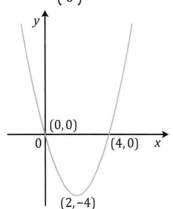

(b) $y = -2f(x)$ is a reflection in the x-axis (because of the negative sign) followed by a stretch parallel to the y-axis with scale factor 2. Note it does not matter in which order these two transformations are applied to the original function.

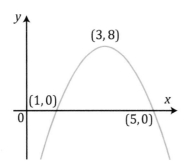

BOOST

Grade

Be on the look out for two transformations made on the same graph.

2 The diagram below shows a sketch of the graph of $y = f(x)$. The graph has a maximum point at (2, 5) and intersects the x-axis at the points (−2, 0) and (6, 0).

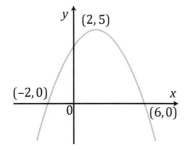

(a) Sketch the graph of $y = f\left(\dfrac{x}{2}\right)$, indicating the coordinates of the stationary point and the coordinates of the points of intersection of the graph with the x-axis. [3]

(b) The diagram below shows a sketch of the graph having **one** of the following equations with an appropriate value of either p, q or r.

$$y = f(x + p), \text{ where } p \text{ is a constant}$$

$$y = f(x) + q, \text{ where } q \text{ is a constant}$$

$$y = rf(x), \text{ where } r \text{ is a constant}$$

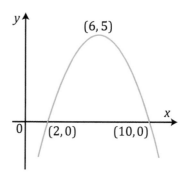

Write down the equation of the graph, together with the value of the corresponding constant.   [2]

. . . .

Answer

2   (a)

The transformation from $y = f(x)$ to $y = f(ax)$ represents a one-way stretch with scale factor $\frac{1}{a}$ parallel to the x-axis.

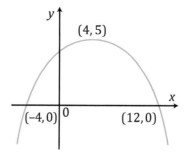

Here the transformation is from $y = f(x)$ to $y = f\left(\frac{x}{2}\right)$. This represents a one-way stretch with scale factor 2 parallel to the x-axis. Each x-coordinate is multiplied by 2 but the y-coordinates remain the same.

$y = f(x + p)$ represents a translation of $\begin{pmatrix} -p \\ 0 \end{pmatrix}$.

$y = f(x) + q$ represents a translation of $\begin{pmatrix} 0 \\ q \end{pmatrix}$.

$y = rf(x)$ represents a one-way stretch with scale factor r parallel to the y-axis

(b) By observing the two graphs, you can see that in the transformed graph the y-coordinates have stayed the same but all the x-coordinates have moved 4 units to the right. This is a translation of $\begin{pmatrix} 4 \\ 0 \end{pmatrix}$

The equation of the translated curve is $y = f(x - 4)$

3   The diagram shows the graph of $y = f(x)$.

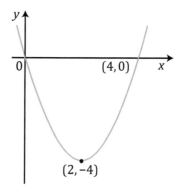

The curve passes through the origin and the point (4, 0), and has a minimum point at (2, −4). Sketch on separate diagrams the graphs of:

(a) $y = -f(x)$ [2]

(b) $y = -f(x - 2)$ [3]

in each case giving the coordinates of the points of intersection of the graph with the x-axis and the coordinates of the stationary point.

> The stationary point is the point where the gradient is zero. In this case it is the coordinates of the maximum or minimum point.

Answer

3 (a) $y = -f(x)$ is a reflection in the x-axis of the graph $y = f(x)$

The points on the x-axis will stay in the same place and the minimum at (2, −4) will be reflected to become a maximum at (2, 4).

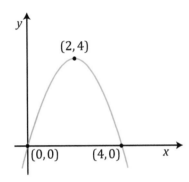

(b) $y = -f(x - 2)$ is translation by $\begin{pmatrix} 2 \\ 0 \end{pmatrix}$ of the graph $y = -f(x)$

The y-coordinates will stay the same but the x-coordinates will be shifted to the right by 2 units.

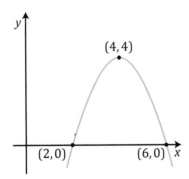

Test yourself

1 Write the following equation using indices:
$$y = 5\sqrt{x} + \frac{45}{x} - 7$$

2 Simplify each of the following without using a calculator:
(a) 5^0
(b) 3^{-2}
(c) $8^{\frac{1}{3}}$
(d) $25^{-\frac{1}{2}}$
(e) $16^{\frac{3}{2}}$

3 Simplify each of the following, expressing your answer in surd form:
(a) $\sqrt{48} + \frac{12}{\sqrt{3}} - \sqrt{27}$

(b) $\frac{2 + \sqrt{5}}{3 + \sqrt{5}}$

4 Find the range of values of k for which the quadratic equation $kx^2 + 5x - 7 = 0$ has no real roots.

5 Solve the inequality $x^2 - 6x + 8 > 0$.

6 Express $5x^2 - 20x + 10$ in the form $y = a(x + b)^2 + c$, where, a, b and c are constants whose values are to be found.

7 Solve the inequality $1 - 3x < x + 7$.

8 Show that the straight line $y = x + 4$ touches the curve $y = x^2 - 7x + 20$ and find the coordinates of the point of contact.

9 Calculate the remainder when $4x^3 + 3x^2 - 3x + 1$ is divided by $x + 1$.

10 (a) Given that $x + 2$ is a factor of $x^3 + 6x^2 + ax + 6$, show that $a = 11$.
(b) Solve the equation $x^3 + 6x^2 + 11x + 6 = 0$.

11 The diagram shows a sketch of the curve $y = x^2 - 4$ and the line joining points Q (2, 0) and R (0, 1). The point P has coordinates (a, b). Write down the three inequalities involving a and b which are such that the point P will be strictly contained **within** the shaded area above, if and only if, all three inequalities are satisfied.

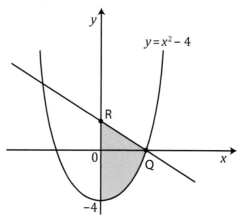

⑫ Sketch graph of the cubic function $y = (x + 4)(x - 2)(x - 7)$ and mark on your graph all the points of intersection of the curve with the x- and y-axes.

⑬ Show that $x^2 - 1.2x - 3.64$ may be expressed in the form $(x - p)^2 - 4$, where p is a constant whose value is to be found.
Hence solve the quadratic equation $x^2 - 1.2x - 3.64 = 0$.

⑭ Given that $k \neq 1$, the quadratic equation $(k - 1)x^2 + kx + k = 0$ has two distinct real roots.
Show that $3x^2 - 4k < 0$.
Find the range of values of k satisfying this inequality.

⑮ Express $4x^2 - 12x + 9$ in the form $a(x + b)^2 + c$, where the values of a, b and c are to be determined. [4]
Hence, sketch the graph of $4x^2 - 12x + 9$, including the coordinates of the stationary point. [3]

⑯ Solve the inequality $x^2 - 2x - 15 \leq 0$.

⑰ The diagram below shows the graph of $y = f(x)$. The graph has a maximum point at $(1, 2)$.

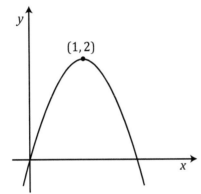

Sketch the following graphs, using a separate set of axes for each graph marking on your graphs the coordinates of the stationary point in each case:
(a) $y = -f(x)$ [2]
(b) $y = 3f(x)$ [2]
(c) $y = f(x - 1)$ [2]
(d) $y = f(2x)$ [2]

Summary

Check you know the following facts:

Indices

Multiplying

$a^m \times a^n = a^{m+n}$

The indices are added together.

Dividing

$a^m \div a^n = a^{m-n}$

The indices are subtracted (i.e. top power minus bottom power).

Power raised to a power

$(a^m)^n = a^{m \times n}$

The indices are multiplied together.

Negative powers

$a^{-m} = \dfrac{1}{a^m}$

Zero power

If $a \neq 0$, $a^0 = 1$

Fractional powers

$a^{\frac{m}{n}} = \sqrt[n]{a^m} = \left(\sqrt[n]{a}\right)^m$

Negative fractional powers

$a^{-\frac{m}{n}} = \dfrac{1}{a^{\frac{m}{n}}} = \dfrac{1}{\sqrt[n]{a^m}}$ or $\dfrac{1}{\left(\sqrt[n]{a}\right)^m}$

Surds

Simple manipulation of surds

$\sqrt{a} \times \sqrt{a} = a$

$\sqrt{a} \times \sqrt{b} = \sqrt{ab}$

$\left(\sqrt{a} + \sqrt{b}\right)\left(\sqrt{a} - \sqrt{b}\right) = a - b$

Rationalisation of surds

We avoid having surds in the denominator and removing them is called rationalising the denominator.

$$\frac{a}{b\sqrt{c}} = \frac{a}{b\sqrt{c}} \times \frac{\sqrt{c}}{\sqrt{c}} = \frac{a\sqrt{c}}{bc}$$

(Here the denominator is rationalised by multiplying the top and bottom by $\sqrt{c}$.)

$$\frac{a}{\sqrt{b} \pm \sqrt{c}} = \frac{a}{\sqrt{b} \pm \sqrt{c}} \times \frac{\sqrt{b} \mp \sqrt{c}}{\sqrt{b} \mp \sqrt{c}} = \frac{a\sqrt{b} \mp a\sqrt{c}}{b - c}$$

(Here the denominator is rationalised by multiplying the top and bottom of the expression by the conjugate of the denominator.)

Quadratic functions and equations

Completing the square

The quadratic expression $ax^2 + bx + c$ can be written in the form $a(x + p)^2 + q$ and this is called completing the square. Completing the square can be used when a quadratic equation cannot be solved by factorisation or you want the find the maximum or minimum value of a quadratic function.

Solving/finding the roots of a quadratic equation

$ax^2 + bx + c$ has roots/solutions given by $x = \dfrac{-b \pm \sqrt{b^2 - 4ac}}{2a}$

Remember this formula as it will **not** be in the formula booklet.

Discriminants of quadratic functions

The discriminant of $ax^2 + bx + c$ is $b^2 - 4ac$

For the equation $ax^2 + bx + c = 0$:

If $b^2 - 4ac > 0$, then there are two real and distinct (i.e. different) roots.

If $b^2 - 4ac = 0$, then there are two real and equal roots

If $b^2 - 4ac < 0$, then there are no real roots

Sketching a quadratic function

First write the equation $ax^2 + bx + c$ in the form $y = a(x + p)^2 + q$

From this equation:

If $a > 0$ the curve will be ∪-shaped.

If $a < 0$ the curve will be ∩-shaped.

The maximum or minimum point will be at $(-p, q)$.

The axis of symmetry will be $x = -p$.

Solving linear inequalities

Solve them in the same way as you would solve ordinary equations but remember to reverse the inequality if you multiply or divide both sides by a negative quantity.

Solving quadratic inequalities

Consider the quadratic function equal to zero and solve to find the values of x where the curve cuts the x-axis.

Draw a sketch of the graph showing the intercepts on the x-axis.

If $ax^2 + bx + c < 0$ then the range of values of x covers the region below the x-axis.

If $ax^2 + bx + c > 0$ then the range of values of x covers the region above the x-axis.

If the inequality includes an equals sign then the range of values of x will include the values where it cuts the x-axis.

Transformations of the graph of $y = f(x)$

Original function	New function	Transformation
$y = f(x)$	$y = f(x) + a$	Translation of a units parallel to the y-axis, i.e. translation of $\begin{pmatrix} 0 \\ a \end{pmatrix}$
$y = f(x)$	$y = f(x + a)$	Translation of a units to the left, parallel to the x-axis, i.e. translation of $\begin{pmatrix} -a \\ 0 \end{pmatrix}$
$y = f(x)$	$y = f(x - a)$	Translation of a units to the right, parallel to the x-axis, i.e. translation of $\begin{pmatrix} a \\ 0 \end{pmatrix}$
$y = f(x)$	$y = -f(x)$	A reflection in the x-axis
$y = f(x)$	$y = af(x)$	One-way stretch with scale factor a parallel to the y-axis
$y = f(x)$	$y = f(ax)$	One-way stretch with scale factor $\frac{1}{a}$ parallel to the x-axis

Original function	New function	
$y = f(x)$	$y = f(x) + a$	
$y = f(x)$	$y = f(x + a)$	
$y = f(x)$	$y = f(x - a)$	
$y = f(x)$	$y = -f(x)$	

Original function	New function	Transformation
$y = f(x)$	$y = af(x)$ E.g. $y = 2f(x)$	
$y = f(x)$	$y = f(ax)$ E.g. $y = f(2x)$	

Algebraic manipulation of polynomials

The remainder theorem

The remainder theorem states:

If a polynomial $f(x)$ is divided by $(x - a)$ the remainder is $f(a)$.

The factor theorem

For a polynomial $f(x)$ if $f(a) = 0$ then $(x - a)$ is a factor of $f(x)$.

3 Coordinate geometry in the (x, y) plane

Introduction

This topic builds on and reinforces your GCSE work on straight line graphs. Make sure you understand gradients and equations of straight lines and solving simultaneous equations before starting this topic. There are quite a few formulae and techniques to learn here but once you have mastered them, this is a relatively easy topic.

This topic covers the following:

3.1 Finding the gradient, length, mid-point and equation, of a line joining two points

3.2 Conditions for two straight lines to be parallel or perpendicular to each other

3.3 The equation of a circle

3.4 Circle properties

3.5 Finding the equation of a tangent to a circle

3.6 Finding where a circle and straight line intersect or meet

3.7 Using the discriminant to identify whether a line and circle intersect and, if so, how many times

3.8 Condition for two circles to touch internally or externally

3.1 Finding the gradient, length, mid-point and equation, of a line joining two points

Finding the gradient of a straight line

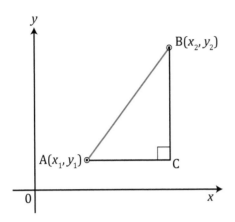

> Be careful with the signs of gradients. A positive gradient slopes upwards with increasing values of x. A negative gradient slopes downwards.

From the above graph length $AC = x_2 - x_1$ and length $BC = y_2 - y_1$

Gradient of line $AB = \dfrac{BC}{AC} = \dfrac{y_2 - y_1}{x_2 - x_1}$

The gradient of the line joining points (x_1, y_1) and (x_2, y_2) is given by:

$$\text{Gradient} = \frac{y_2 - y_1}{x_2 - x_1}$$

> You need to remember this formula as it will not be given in the formula booklet.

For example the gradient of the straight line joining the points A(−3, 2) and B(1, 6)

is $\dfrac{6 - 2}{1 - (-3)} = \dfrac{4}{4} = 1$

Finding the length of a straight line joining two points

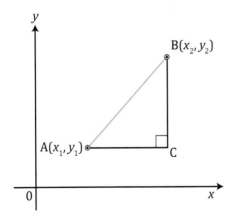

Now $AC = x_2 - x_1$ and length $BC = y_2 - y_1$

By Pythagoras' Theorem $AB^2 = AC^2 + BC^2$

So $AB^2 = (x_2 - x_1)^2 + (y_2 - y_1)^2$

$$AB = \sqrt{(x_2 - x_1)^2 + (y_2 - y_1)^2}$$

The length of a straight line joining the two points (x_1, y_1) and (x_2, y_2) is given by:

Remember this formula.

$$\boxed{\sqrt{(x_2 - x_1)^2 + (y_2 - y_1)^2}}$$

For example the length of the straight line joining the points A $(-3, -1)$ and B $(1, 2)$ is $\sqrt{(1 - (-3))^2 + (2 - (-1))^2} = \sqrt{16 + 9} = \sqrt{25} = 5$ units

Finding the mid-point of a straight line joining two points

The mid-point of a line joining the points (x_1, y_1) and (x_2, y_2) is given by:

Remember this formula.

$$\boxed{\left(\frac{x_1 + x_2}{2}, \frac{y_1 + y_2}{2} \right)}$$

For example the mid-point of the line joining the two points with coordinates $(2, 6)$ and $(8, 4)$ is

$$\left(\frac{2 + 8}{2}, \frac{6 + 4}{2} \right) = (5, 5)$$

Finding the equation of a straight line

To find the equation of a straight line you need to know the gradient of the line (m) and the coordinates of a point (x_1, y_1) that lies on the line.

The equation of a straight line with gradient m and which passes through a point (x_1, y_1) is given by:

Remember this formula.

$$\boxed{y - y_1 = m(x - x_1)}$$

For example, the equation of the straight line with a gradient of 2 and passing through the point $(2, 5)$ is:

$$y - 5 = 2(x - 2)$$
$$y - 5 = 2x - 4$$

So $\qquad\qquad y = 2x + 1 \quad$ or $\quad 2x - y + 1 = 0$

Notice that the first equation is in the form $y = mx + c$, so you can immediately see that the gradient, $m = 2$ and the intercept on the y-axis, $c = +1$.

The second form gives the equation as $ax + by + c = 0$.

Which form you use depends on whether you are asked for a particular form in the question. You can give the equation of a straight line in either form, if a form is not specified in the question.

Examples

1   Find the equation of the line L, having gradient 3 and passing through the point $(2, 3)$.

Answer

1   $y - y_1 = m(x - x_1)$ where $m = 3$ and $(x_1, y_1) = (2, 3)$.

$y - 3 = 3(x - 2)$

$y - 3 = 3x - 6$

$y = 3x - 3$ This equation is in the form $y = mx + c$, where m is the gradient and c is the intercept on the y-axis.

> Write down the general equation for a straight line and then substitute values into it for m, x_1 and y_1.

---

2   Find the equation of the line in the form $ax + by + c = 0$ that has a gradient of 2 and passes through the point $(-1, 0)$.

Answer

2   $y - y_1 = m(x - x_1)$ where $m = 2$ and $(x_1, y_1) = (-1, 0)$.

$y - 0 = 2(x - (-1))$

$y = 2(x + 1)$

$y = 2x + 2$

$2x - y + 2 = 0$

> Remember to give the equation in the format asked for in the question.

---

3   Find the length of the line joining the two points $(-1, -2)$ and $(4, 10)$.

Answer

3   Using the formula for the distance between two points.

The length of a straight line joining the two points (x_1, y_1) and (x_2, y_2) is given by:

$$\sqrt{(x_2 - x_1)^2 + (y_2 - y_1)^2}$$

Substituting the coordinates $(-1, -2)$ and $(4, 10)$ into this gives

Length $= \sqrt{(4 - (-1))^2 + (10 - (-2))^2} = \sqrt{25 + 144} = \sqrt{169} = 13$ units

> Be careful substituting negative numbers into this formula. It is best to add brackets to emphasise the negative numbers.

3.2 Conditions for two straight lines to be parallel or perpendicular to each other

Condition for two straight lines to be parallel to each other

> For two lines to be parallel to each other, they must have the same gradient.

For example, the equation of the line that is parallel to the line $y = 3x - 2$, but intersects the y-axis at $y = 2$ is:

$y = 3x + 2$ as $m = 3$ and $c = 2$ (i.e. using the equation $y = mx + c$).

Condition for two straight lines to be perpendicular to each other

> When two lines are perpendicular to each other (i.e. they make an angle of 90°), the product of their gradients is –1.

If one line has a gradient m_1 and the other a gradient of m_2 then

$$m_1 m_2 = -1$$

For example, if a straight line has gradient $-\frac{1}{3}$ then the gradient of the line perpendicular to this is given by:

$$\left(-\frac{1}{3}\right)m_2 = -1, \text{ hence gradient } m_2 = 3$$

Examples

1 Find the equation of line L_1 which passes through the point (1, 2) and is parallel to the line L_2 which has the equation $2x - y + 1 = 0$.

. . . .

Answer

1   First find the equation of the line L_2 in the form $y = mx + c$

$2x - y + 1 = 0$

So $y = 2x + 1$

Hence the gradient of $L_2 = 2$.

Since lines L_1 and L_2 are parallel, they both have the same gradient of 2. Finding the equation of line L_1

$y - y_1 = m(x - x_1)$ where $m = 2$ and $(x_1, y_1) = (1, 2)$.

$y - 2 = 2(x - 1)$

$y = 2x$ (or $y - 2x = 0$)

> Comparing this equation with $y = mx + c$ gives the gradient, $m = 2$.

> **BOOST**
> **Grade** ⬆⬆⬆⬆
> Always check to see if a form for the equation of a straight line is given in the question. If the form is not specified, then either equation specified here is acceptable.

2 The points A, B and C have coordinates (1, 1), (3, 3) and (6, 0) respectively.

(a) Find the gradients of lines AB and BC.

(b) Prove that lines AB and BC are perpendicular to each other.

. . . .

Answer

2   (a) Gradient of AB $= \dfrac{3 - 1}{3 - 1} = 1$

Gradient of BC $= \dfrac{0 - 3}{6 - 3} = -1$

(b) Product of the gradients $= (1)(-1) = -1$.

As $m_1 m_2 = -1$, AB and BC are perpendicular to each other.

> Both gradients are found using the formula:
> Gradient $= \dfrac{y_2 - y_1}{x_2 - x_1}$

3 The points A, B, C have coordinates $(-11, 10)$, $(-5, 12)$, $(3, 8)$ respectively.

The line L_1 passes through the point A and is **parallel** to BC.

The line L_2 passes through the point C and is **perpendicular** to BC.

(a) Find the gradient of BC. [2]

(b) (i) Show that L_1 has equation: $x + 2y - 9 = 0$.

    (ii) Find the equation of L_2. [6]

(c) The lines L_1 and L_2 intersect at the point D.

    (i) Show that D has coordinates $(1, 4)$.

    (ii) Find the length of BD.

    (iii) Find the coordinates of the mid-point of BD. [6]

. . . .

Answer

3 (a) Gradient of BC $= \dfrac{8 - 12}{3 - (-5)} = \dfrac{-4}{8} = -\dfrac{1}{2}$

(b) (i) Gradient of line $L_1 = -\dfrac{1}{2}$ as lines L_1 and BC are parallel.

$m = -\dfrac{1}{2}$ and $(x_1, y_1) = (-11, 10)$

The equation of L_1 is given by

$$y - y_1 = m(x - x_1)$$

$$y - 10 = -\frac{1}{2}(x - (-11))$$

$$2y - 20 = -x - 11$$

$$x + 2y - 9 = 0$$

(ii) Gradient of line $L_2 = 2$

The equation of L_2 is given by $y - y_1 = m(x - x_1)$ where $m = 2$ and $(x_1, y_1) = (3, 8)$.

$$y - 8 = 2(x - 3)$$

$$y - 8 = 2x - 6$$

$$2x - y + 2 = 0$$

Here we use the fact that the product of the gradients of perpendicular lines is –1.

(c) (i) Solving the equations of lines L_1 and L_2 simultaneously to find the point of intersection:

$$x + 2y - 9 = 0 \tag{1}$$

$$2x - y + 2 = 0 \tag{2}$$

Multiplying equation (1) by 2:

$$2x + 4y - 18 = 0$$

$$2x - y + 2 = 0$$

BOOST

Grade ⇧⇧⇧⇧

Being able to solve simultaneous equations is assumed at AS level. You may need to go back to your GCSE work to check you can do them.

Subtracting these two equations gives:

$$5y - 20 = 0$$
$$y = 4$$

Substituting $y = 4$ into equation (1) gives:

$$x + 8 - 9 = 0$$
$$x - 1 = 0$$
$$x = 1$$

Checking by substituting the values of x and y into equation (2)

$$LHS = 2x - y + 2$$
$$= 2(1) - 4 + 2 = 0 = RHS$$

Hence D is the point (1, 4)

> Always check the values for x and y by substituting them into the equation that you have not used already for the substitution.

(ii) The length of a straight line joining the two points (x_1, y_1) and (x_2, y_2) is given by:

$$\sqrt{(x_2 - x_1)^2 + (y_2 - y_1)^2}$$

Length of BD $= \sqrt{(-5 - 1)^2 + (12 - 4)^2}$

$$= \sqrt{36 + 64}$$
$$= \sqrt{100}$$
$$= 10$$

(iii) The mid-point of a line joining the points (x_1, y_1) and (x_2, y_2) is given by:

$$\left(\frac{x_1 + x_2}{2}, \frac{y_1 + y_2}{2}\right)$$

Mid-point of BD $= \left(\frac{-5 + 1}{2}, \frac{12 + 4}{2}\right) = (-2, 8)$

---

4 A line passes through the points A (1, −1) and B (3, 4).

(a) Find the gradient of line AB. [2]

(b) Find the coordinates of C, the mid-point of AB. [2]

(c) The line L is perpendicular to line AB and passes through the point C. Find the equation of line L. [3]

· ·

Answer

4 (a) Gradient $= \dfrac{y_2 - y_1}{x_2 - x_1} = \dfrac{4 - (-1)}{3 - 1} = \dfrac{5}{2}$

(b) The mid-point of a line joining the points (x_1, y_1) and (x_2, y_2) is given by:

$$\left(\frac{x_1 + x_2}{2}, \frac{y_1 + y_2}{2}\right)$$

Hence mid-point of AB $= \left(\frac{1 + 3}{2}, \frac{-1 + 4}{2}\right) = \left(2, \frac{3}{2}\right)$

(c) The product of the gradients of perpendicular lines is –1.

Hence $m\left(\dfrac{5}{2}\right) = -1$

Giving $m = -\dfrac{2}{5}$

Equation of straight line L having gradient $-\dfrac{2}{5}$ and passing through the point $\left(2, \dfrac{3}{2}\right)$ is: $\quad y - \dfrac{3}{2} = -\dfrac{2}{5}(x - 2)$

Multiplying through by 10 gives:

$$10y - 15 = -4(x - 2)$$

$$10y - 15 = -4x + 8$$

$$4x + 10y - 23 = 0$$

> Where there are fractions like this, multiply both sides by the lowest common denominator.

---

5 The points A, B, C, D have coordinates (–4, 4), (–1, 3), (0, 1), (k, 0) respectively. The straight line CD is parallel to the straight line AB.

  (a) Find the gradient of AB.   [2]

  (b) Find the gradient of CD and hence find the value of the constant k.   [3]

  (c) Line L is perpendicular to CD and passes through point C. Find the equation of line L in the form $ax + by + c = 0$.   [2]

....

Answer

5   (a) Gradient of AB $= \dfrac{y_2 - y_1}{x_2 - x_1} = \dfrac{3 - 4}{-1 - (-4)} = -\dfrac{1}{3}$

  (b) Gradient of CD $= \dfrac{y_2 - y_1}{x_2 - x_1} = \dfrac{0 - 1}{k - 0} = -\dfrac{1}{k}$

    As line CD is parallel to AB the gradients are equal.

    Hence, $\quad -\dfrac{1}{3} = -\dfrac{1}{k}$

    Giving $\quad k = 3$

> The gradients are equated here.

  (c) As line L is perpendicular to CD the product of their gradients is –1.

    Hence $\quad \left(-\dfrac{1}{3}\right)m_2 = -1$

    Giving gradient of L = 3

    Equation of line L is:

$$y - 1 = 3(x - 0)$$

$$y = 3x + 1$$

    Hence $\quad 3x - y + 1 = 0$

BOOST

Grade ⇧⇧⇧⇧

Leaving the equation in the form $y = mx + c$ (i.e. $y = 3x + 1$) would cost you a mark here as the question asks that the equation be given in the form $ax + by + c = 0$.

3.3 The equation of a circle

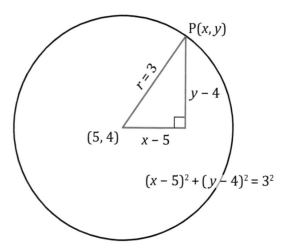

The formats for the equations of circles will not be included in the formula booklet.

The equation of a circle can be written in the form:

$$(x - a)^2 + (y - b)^2 = r^2$$

A circle having the above equation will have centre (a, b) and radius r.
There is the following alternative form for the equation of a circle:

$$x^2 + y^2 + 2gx + 2fy + c = 0$$

A circle having the above equation will have centre $(-g, -f)$ and radius given by:

$$r = \sqrt{g^2 + f^2 - c}$$

Note: You have to remember this alternative form for the equation as well as be able to work out the centre and radius. Remembering this is hard, so there is an alternative method which involves completing the square and this method is shown in Example 3.

Examples

1 Find the coordinates of the centre and the radius of the circle having the equation:

$$(x - 7)^2 + (x + 3)^2 = 36$$

. . . .

Answer

1   Comparing the equation $(x - 7)^2 + (x + 3)^2 = 36$ with the equation for the circle

$$(x - a)^2 + (y - b)^2 = r^2$$

This gives $a = 7$ and $b = -3$ so coordinates of the centre are $(7, -3)$.

$r^2 = 36$, giving radius $r = \sqrt{36} = 6$

2 The circle C has centre A and equation $x^2 + y^2 - 2x + 6y - 6 = 0$.
Write down the coordinates of A and find the radius of C.

Answer

2 Comparing the equation $x^2 + y^2 - 2x + 6y - 6 = 0$ with the equation
$x^2 + y^2 + 2gx + 2fy + c = 0$, we can see $g = -1, f = 3, c = -6$.

Centre A has coordinates $(-g, -f) = (1, -3)$

Radius $= \sqrt{g^2 + f^2 - c} = \sqrt{(-1)^2 + (3)^2 + 6} = \sqrt{16} = 4$

3 The circle C has centre A and equation $x^2 + y^2 - 4x + 2y - 11 = 0$
Find the coordinates of A and the radius of C.

Answer

3 The equation for C can be written as

$x^2 - 4x + y^2 + 2y - 11 = 0$

Completing the square means $x^2 - 4x = (x - 2)^2 - 4$

Similarly $y^2 + 2y = (y + 1)^2 - 1$

Hence, equation of C can be written as:

$(x - 2)^2 - 4 + (y + 1)^2 - 1 - 11 = 0$

$(x - 2)^2 + (y + 1)^2 - 4 - 1 - 11 = 0$

$(x - 2)^2 + (y + 1)^2 = 16$

Comparing this with the equation of the circle $(x - a)^2 + (y - b)^2 = r^2$ gives the
coordinates of the centre A as $(2, -1)$ and radius $= 4$.

> Completing the square
> is a good way of finding
> the centre and radius
> because you do not need
> to remember the formula
> involving f and g.

3.4 Circle properties

There are a number of circle properties you need to know about.

1 The angle in a semicircle is always a right angle.

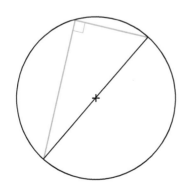

2   The perpendicular from the centre of a circle to a chord bisects the chord.

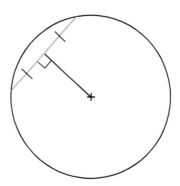

3   The tangent to a circle at a point makes a right angle with the radius of the circle at the same point.

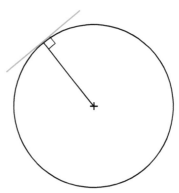

3.5 Finding the equation of a tangent to a circle

If you know the coordinates of the point where the tangent touches the circle and the coordinates of the centre of the circle, then you can find the gradient of the line joining these two points using the formula:

$$\text{Gradient} = \frac{y_2 - y_1}{x_2 - x_1}$$

You would then use this gradient to work out the gradient of the tangent as these two lines are perpendicular to each other. If one line has a gradient m_1 and the other a gradient of m_2 then as the lines are perpendicular $m_1 m_1 = -1$

You would then use the coordinates of the point where the tangent touched the circle and the gradient of the tangent and substitute them into following formula to give the equation of the tangent.

$$y - y_1 = m(x - x_1)$$

The following example will help explain the method.

Example

1   Circle C has centre A and equation $x^2 + y^2 - 4x + 2y - 20 = 0$.

(a)   Find the coordinates of the centre A and the radius of C.   [3]

(b) The point P has coordinates (5, 3) and lies on circle C. Find the equation of the tangent to C at P. [4]

Answer

1 (a) We will use the method of completing the square here to work out the coordinates of the centre A and the radius of the circle C.

$$x^2 + y^2 - 4x + 2y - 20 = 0$$

$$(x - 2)^2 + (y + 1)^2 - 4 - 1 - 20 = 0$$

$$(x - 2)^2 + (y + 1)^2 = 25$$

$$(x - 2)^2 + (y + 1)^2 = 5^2$$

Hence coordinates of the centre A are (2, −1) and radius is 5.

> Completing the square is used here but you could of course use the alternative method involving the formula. You would need to remember the formula and how to use it as it is not in the formula booklet.
>   See page 25.

(b) Gradient of the line joining the centre of the circle A (2, −1) to point P (5, 3) is given by:

Gradient of AB $= \dfrac{y_2 - y_1}{x_2 - x_1} = \dfrac{3 - (-1)}{5 - 2} = \dfrac{4}{3}$

Line AP is a radius of the circle. The tangent at point P will be perpendicular to the radius AP.

For perpendicular lines, the product of the gradients = −1

Hence $m \times \dfrac{4}{3} = -1$

Gradient of tangent $m = \dfrac{-3}{4}$

Equation of the tangent having gradient $m = -\dfrac{3}{4}$ and passing through the point (5, 3) is

$$y - 3 = -\dfrac{3}{4}(x - 5)$$

$$4y - 12 = -3x + 15$$

$$3x + 4y - 27 = 0$$

> The formula for a straight line is used here.
>
> The formula for the equation of a straight line having gradient m and passing through the point (x_1, y_1) is
> $y - y_1 = m(x - x_1)$.

3.6 Finding where a circle and straight line intersect or meet

There are two ways in which a straight line can intersect or meet a circle:

1 The line and circle can intersect in two places like this:

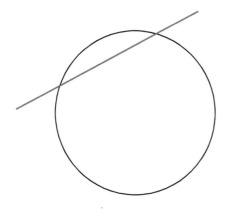

2 The line and circle can meet in one place. This means the straight line becomes a tangent to the circle and also makes a right angle with the radius.

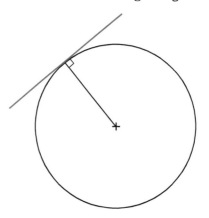

To find the coordinates of intersection or meeting

To find the coordinates you need to know the equation of the circle and the equation of the straight line. These are then solved simultaneously. You can use the straight line equation to find x in terms of y or y in terms of x. You then substitute this into the equation of the circle and then solve the resulting equation. Sometimes there will be two different roots (i.e. solutions), which means the circle and line cut in two places. Sometimes there will be two equal roots which means the circle and line meet in one place, i.e. the line is a tangent to the circle.

If there are no real roots to the equation, it means the line and circle do not intersect.

> If there are two solutions when two equations of lines/curves are solved, it means they intersect in two places. If there is only one solution they only intersect or touch in one place. It is also possible that there are no real solutions which means they neither touch nor intersect.

Example

1 A circle C has the equation $x^2 + y^2 + 2x - 12y + 12 = 0$

The line with equation $x + y = 4$ intersects the circle at two points P and Q. Find the coordinates of P and Q.

. . . .

Answer

1   To find the points of intersection we solve the two equations simultaneously.

$$x + y = 4$$

So $y = 4 - x$

Substituting $y = 4 - x$ into the equation of the circle, gives

$$x^2 + (4 - x)^2 + 2x - 12(4 - x) + 12 = 0$$
$$x^2 + 16 - 8x + x^2 + 2x - 48 + 12x + 12 = 0$$
$$2x^2 + 6x - 20 = 0$$

Dividing by two gives $x^2 + 3x - 10 = 0$

Factorising gives $(x + 5)(x - 2) = 0$

Solving gives $x = -5$ or 2

> Always look at a quadratic to see if all the terms can be divided by the same number. This will make the factorisation easier.

These two x-coordinates are substituted into the equation of the line to find the corresponding y-coordinates.

When $x = -5, y = 4 - (-5) = 9$

When $x = 2, y = 4 - 2 = 2$

Hence coordinates of points of intersection P and Q are $(-5, 9)$ and $(2, 2)$.

3.7 Using the discriminant to identify whether a line and circle intersect and, if so, how many times

If the circle and line do not meet or intersect, the resulting quadratic equation, when the two equations are solved simultaneously, will have no real roots.

To prove that there are no real roots of a quadratic equation in the form $ax^2 + bx + c = 0$ we can show that the discriminant $b^2 - 4ac < 0$.

> Using the discriminant you can prove whether curves and lines cut/touch and in how many places.

Note also that:

- If $b^2 - 4ac > 0$ there are two real and distinct roots, meaning the circle and line intersect in two places.

- If $b^2 - 4ac = 0$ there are two real and equal roots (i.e. only one solution), meaning the circle and line meet in one place. The line is therefore a tangent to the circle.

3.8 Condition for two circles to touch internally or externally

When two circles touch externally it means that one of the circles is outside the other and they touch at a single point. When circles touch internally, one of the circles is inside the other.

Circles touching externally at a point

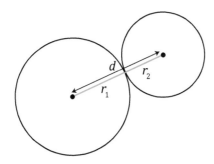

When two circles touch externally at a point, the distance between the centres of the circles must equal the sum of the radii of the two circles.

If the distance between the centres is d and the radii of the two circles are r_1 and r_2 then if the circles touch externally at a point:

$$d = r_1 + r_2$$

Step by

A circle has equation $x^2 + y^2 - 2x - 10y + 18 = 0$ and centre C.

(a) Find the equation of the chord of the circle with a mid-point of P (2, 6)

(b) Prove that the line $x + y = 10$ is a tangent to circle and find the coordinates of its point of contact D.

(c) Prove that points C, D and P all lie in a straight line.

Steps to take

1 First use the equation of the circle to find the centre and the radius.

2 Draw a quick sketch of the circle and the mid-point of the chord. It is worth the small amount of time it takes.

3 Look for any circle properties which might help you.

4 If a line is a tangent to a curve then when they are solved simultaneously the resulting equation will only have real and equal roots.

5 To prove points lie in a straight line, you can find the gradients of the line segments which should be the same. You cannot have lines with the same gradient, passing through the same point, that are not on the same straight line.

Answer

(a) $x^2 + y^2 - 2x - 10y + 18 = 0$

$(x - 1)^2 + (y - 5)^2 - 1 - 25 + 18 = 0$

$(x - 1)^2 + (y - 5)^2 = 8$

Hence centre C is at (1, 5) and radius is $\sqrt{8}$ or $2\sqrt{2}$.

The sketch is now drawn

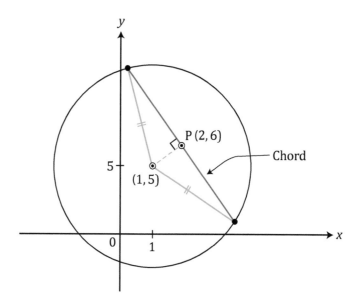

Gradient of line joining points C (1, 5) and P (2, 6) = $\frac{6-5}{2-1}$ = 1

Hence, gradient of the chord = −1 (i.e. using $m_1 m_2 = -1$)

Equation of chord with gradient −1 and passing through P (2, 6) is

$$y - 6 = -1(x - 2)$$

$$y = -x + 8$$

(b) $x + y = 10$ so $y = 10 - x$

Substituting this value of y^2 into the equation of a circle we have:

$$x^2 + (10 - x)^2 - 2x - 10(10 - x) + 18 = 0$$

$$x^2 + 100 - 20x + x^2 - 2x - 100 + 10x + 18 = 0$$

$$2x^2 - 12x + 18 = 0$$

$$x^2 - 6x + 9 = 0$$

$$(x - 3)(x - 3) = 0 \qquad \text{so } x = 3$$

As there are two equal roots the line meets the circle at one point, so the line is a tangent.

When $x = 3$, $x + y = 10$, so $y = 7$

D is the point (3, 7)

(c) Gradient of line PD = $\frac{7-6}{3-2}$ = 1

Gradient of line CP = 1 (already found in part (a))

Now as the gradients are equal and both lines pass through point P, they must lie in a straight line.

> The line from the centre of the circle to point P is at right angles to the chord. If the gradient of this line is found then the gradient of the chord can be found and hence its equation.

> Substitute the x-coordinate into the equation of the line (because it is simpler than the curve) to find the corresponding y-coordinate.

Example

1 Circle 1 has centre A (−2, 1) and radius 5. Circle C_2 has centre B (10, 6) and radius r. If the circles C_1 and C_2 touch externally, find the value of r

. . . .

Answer

1   If d is the distance between the centres A and B, then

$$d = \sqrt{(x_2 - x_1)^2 + (y_2 - y_1)^2}$$

$$= \sqrt{(10 - (-2))^2 + (6 - 1)^2}$$

$$= \sqrt{144 + 25}$$

$$= \sqrt{169}$$

$$= 13$$

As the circles touch externally, the sum of the radii must equal the distance between the centres.

Hence, $13 = r + 5$, giving $r = 8$

Circles touching internally at a point

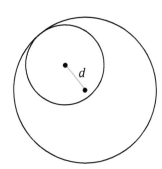

When two circles touch internally at a point, the distance between the centres of the circles equals the difference of the radii of the two circles.

If the distance between the centres is d and the radii of the two circles are r_1 and r_2 then if the circles touch internally at a point:

$$d = r_1 - r_2$$

Example

1   Two circles C_1 and C_2 touch internally at a point. Circle C_1 has the following equation:
$$x^2 + y^2 - 4x - 4y - 1 = 0$$

If circle C_2 has centre (2, 1) find the radius of this circle.

. . . .

Answer

1   $x^2 + y^2 - 4x - 4y - 1 = 0$

$$(x - 2)^2 + (y - 2)^2 - 4 - 4 - 1 = 0$$

$$(x - 2)^2 + (y - 2)^2 = 9$$

Hence circle C_1 has centre (2, 2) and radius 3

Distance between the centres of both circles

$$d = \sqrt{(x_2 - x_1)^2 + (y_2 - y_1)^2}$$
$$= \sqrt{(2 - 2)^2 + (2 - 1)^2}$$
$$= 1$$

Now if the circles touch internally

$$d = r_1 - r_2$$
$$1 = 3 - r_2$$
$$r_2 = 2$$

Hence the radius of circle $C_2 = 2$

Examples

1   The circle C has centre A and equation $x^2 + y^2 - 2x + 6y - 15 = 0$

(a) (i)   Write down the coordinates of A.

(ii) The point P has coordinates (4, −7) and lies on C. Find the equation of the tangent to C at P.   [5]

(b) The line L has equation $y = x + 4$. Show that L and C do not intersect. [4]

Answer

1 (a) (i) $x^2 + y^2 - 2x + 6y - 15 = 0$

Completing the square gives the following:

$$(x - 1)^2 + (y + 3)^2 - 1 - 9 - 15 = 0$$

$$(x - 1)^2 + (y + 3)^2 = 25$$

Comparing this with the equation of the circle $(x - a)^2 + (y - b)^2 = r^2$ gives the coordinates of centre A as $(1, -3)$ and radius = 5.

> You could alternatively use the formula to work out the coordinates of the centre and the radius of the circle.

(ii) The line joining P and A will be a radius of the circle C.

Gradient of line joining P $(4, -7)$ and A $(1, -3)$ = $\dfrac{-7 - (-3)}{4 - 1} = \dfrac{-4}{3}$

The radius AP will be the normal to the circle at P. The product of the gradients of the tangent and normal will be -1.

> A normal and a tangent make an angle of 90° to each other.

Gradient of tangent at P = $\dfrac{3}{4}$ (i.e. using $m_1 m_2 = -1$)

Equation of the tangent at C having gradient $\dfrac{3}{4}$ and passing through the point $(4, -7)$ is:

$$y - (-7) = \frac{3}{4}(x - 4)$$

$$4y + 28 = 3x - 12$$

$$3x - 4y - 40 = 0$$

> Remember: The equation of a straight line having gradient m and passing through the point (x_1, y_1) is given by: $y - y_1 = m(x - x_1)$

(b) Substituting $y = x + 4$ into the equation of the circle gives:

$$x^2 + (x + 4)^2 - 2x + 6(x + 4) - 15 = 0$$

Multiplying out the brackets and simplifying gives:

$$2x^2 + 12x + 25 = 0$$

Comparing this equation with $ax^2 + bx + c$, gives $a = 2$, $b = 12$ and $c = 25$.

Checking the roots of this equation:

As $b^2 - 4ac < 0$, there are no real roots which means the circle and line do not intersect.

> To find the points of intersection of a circle and a line you solve the two equations simultaneously.
>
> If there is one solution, then the line meets the circle at one point (i.e. it is a tangent).
>
> If there are two solutions, it cuts the circle in two places.
>
> If there are no real solutions the circle and line do not intersect or meet.

2 The points A, B, C, D have coordinates $(-7, 4)$, $(3, -1)$, $(6, 1)$, $(k, -15)$ respectively.

(a) Find the gradient of AB. [2]

(b) Find the equation of AB and simplify your answer. [3]

(c) Find the length of AB. [2]

(d) The point E is the mid-point of AB. Find the coordinates of E. [2]

(e) Given that CD is perpendicular to AB, find the value of the constant k. [4]

Watch the signs when using this formula. Add brackets to the negative coordinates to emphasise them.

Answer

2   (a)   Gradient of AB $= \dfrac{y_2 - y_1}{x_2 - x_1} = \dfrac{4 - (-1)}{-7 - 3} = -\dfrac{1}{2}$

(b)   Equation of straight line which passes through $(-7, 4)$ and has gradient $-\dfrac{1}{2}$ is given by:

$$y - y_1 = m(x - x_1) \text{ where } m = -\dfrac{1}{2} \text{ and } (x_1, y_1) = (-7, 4)$$

$$y - 4 = -\dfrac{1}{2}(x - (-7))$$

$$2y - 8 = -x - 7$$

$$x + 2y - 1 = 0$$

(c)   The length of a straight line joining the two points (x_1, y_1) and (x_2, y_2) is given by:

$$\sqrt{(x_2 - x_1)^2 + (y_2 - y_1)^2}$$

Substituting the coordinates $(-7, 4)$ and $(3, -1)$ into this gives

Length AB $= \sqrt{(3 - (-7))^2 + (-1 - 4)^2} = \sqrt{100 + 25} = \sqrt{125} = 5\sqrt{5}$ units

(d)   The mid-point of a line joining the points (x_1, y_1) and (x_2, y_2) is given by:

$$\left(\dfrac{x_1 + x_2}{2}, \dfrac{y_1 + y_2}{2} \right)$$

Hence mid-point E of AB $= \left(\dfrac{-7 + 3}{2}, \dfrac{4 + (-1)}{2} \right) = \left(-2, \dfrac{3}{2} \right)$

(e)   If CD is perpendicular to AB then the product of the gradients equals -1.

$$m_1 m_2 = -1$$

$$\left(-\dfrac{1}{2} \right) m_2 = -1$$

Giving gradient of CD $= 2$

Finding the gradients using the coordinates of C $(6, 1)$ and D $(k, -15)$ gives

$$\text{Gradient of CD} = \dfrac{y_2 - y_1}{x_2 - x_1} = \dfrac{-15 - 1}{k - 6}$$

The gradient of CD is 2 so an equation can be formed.

Hence $\dfrac{-15 - 1}{k - 6} = 2$

$$-16 = 2(k - 6)$$

$$-16 = 2k - 12$$

$$k = -2$$

---

3   The points A $(0, 2)$, B $(-2, 8)$, C $(20, 12)$ are the vertices of the triangle ABC.

The point D is the mid-point of AB.

(a)   Show that CD is perpendicular to AB.

(b)   Find the exact value of tan CAB.

(c)   Write down the geometrical name for the triangle ABC.

Answer

3 (a)

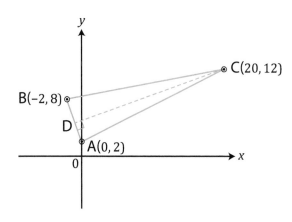

BOOST

Grade ⇧⇧⇧⇧

Sometimes it is worth spending time drawing a diagram so you can see the positions of the points and also better understand the problem.

Mid-point D of AB = $\left(\dfrac{0 + (-2)}{2}, \dfrac{2 + 8}{2}\right)$ = $(-1, 5)$

Gradient of CD = $\dfrac{12 - 5}{20 - (-1)} = \dfrac{7}{21} = \dfrac{1}{3}$

Gradient of AB = $\dfrac{8 - 2}{-2 - 0} = \dfrac{6}{-2} = -3$

Product of the gradients of these two lines = $\dfrac{1}{3} \times (-3) = -1$

Hence the lines are perpendicular to each other.

(b) Tan CAB = $\dfrac{CD}{AD}$

> We need to find the lengths of sides CD and AD.

CD = $\sqrt{(x_2 - x_1)^2 + (y_2 - y_1)^2}$

= $\sqrt{(20 - (-1))^2 + (12 - 5)^2}$

= $\sqrt{441 + 49}$

= $\sqrt{490}$

= $7\sqrt{10}$

AD = $\sqrt{(0 - (-1))^2 + (2 - 5)^2}$

= $\sqrt{10}$

tan CAB = $\dfrac{CD}{AD} = 7\dfrac{\sqrt{10}}{\sqrt{10}} = 7$

(c) Triangle is isosceles (as the perpendicular bisects the base forming two equal right-angled triangles).

---

4 The circle C has centre A and equation $x^2 + y^2 - 4x + 6y = 3$

(a) Write down the coordinates of A and find the radius of C. [3]

(b) A straight line has equation $y = 4x - 7$. This straight line intersects the circle C at two points. Find the coordinates of these two points. [4]

Answer

4   (a)   Comparing the equation $x^2 + y^2 - 4x + 6y = 3$ with the equation $x^2 + y^2 + 2gx + 2fy + c = 0$, we can see $g = -2, f = 3, c = -3$.

Centre A has coordinates $(-g, -f) = (2, -3)$

Radius $= \sqrt{g^2 + f^2 - c} = \sqrt{(-2)^2 + (3)^2 + 3} = \sqrt{16} = 4$

(b)   Substituting $y = 4x - 7$ into the equation for the circle gives:

$$x^2 + (4x - 7)^2 - 4x + 6(4x - 7) = 3$$

$$x^2 + 16x^2 - 56x + 49 - 4x + 24x - 42 = 3$$

$$17x^2 - 36x + 4 = 0$$

$$(17x - 2)(x - 2) = 0$$

$$x = \frac{17}{2} \text{ or } x = 2$$

The *x*-coordinates are substituted into the equation of the straight line to find the *y*-coordinates of the points of intersection.

When $x = \frac{17}{2}$, $y = 4\left(\frac{17}{2}\right) - 7 = 27$

When $x = 2$, $y = 4(2) - 7 = 1$

Hence the curve intersects the straight line at the points $\left(\frac{17}{2}, 27\right)$ and $(2, 1)$.

---

5   The circle C has centre A and radius r. The points P $(0, 5)$ and Q $(8, -1)$ are at either end of a diameter of C.

(a)   (i)   Write down the coordinates of A.

(ii)   Show that $r = 5$.

(iii)   Write down the equation of C.

[4]

(b)   Verify that the point R $(7, 6)$ lies on C.   [2]

(c)   Find the equation of the tangent at point R.   [3]

Answer

5   (a)   (i)   A is the mid-point of PQ. Hence coordinates of A are

The formula for the mid-point:
$$\left(\frac{x_1 + x_2}{2}, \frac{y_1 + y_2}{2}\right)$$
is used here.

$$\left(\frac{0 + 8}{2}, \frac{5 + (-1)}{2}\right) = (4, 2)$$

(ii)   The length of the straight line joining the two points A $(4, 2)$ and P $(0, 5)$ is given by:

Distance AP $= r = \sqrt{(x_2 - x_1)^2 + (y_2 - y_1)^2}$

$= \sqrt{(0 - 4)^2 + (5 - 2)^2}$

$= \sqrt{16 + 9}$

$= \sqrt{25}$

$= 5$

The equation of a circle having centre (a, b) and radius r is given by
$$(x - a)^2 + (y - b)^2 = r^2$$

(iii)   Equation of the circle is $(x - 4)^2 + (y - 2)^2 = 25$

(b) Substituting the coordinates of R (7, 6) into the LHS gives:

$$LHS = (7 - 4)^2 + (6 - 2)^2 = 9 + 16 = 25 = 5^2 = RHS$$

So the coordinates of R lie on the circle.

(c) Gradient of line AR $= \dfrac{6-2}{7-4} = \dfrac{4}{3}$ AR is a radius, so it is perpendicular to the tangent at R.

Gradient of tangent $= -\dfrac{3}{4}$

Equation of tangent is

$$y - 6 = -\frac{3}{4}\left(x - 7\right)$$

$$4y - 24 = -3x + 21$$

$$3x + 4y - 45 = 0$$

> Here you prove that the left-hand side of the equation, with the coordinates of the point entered for x and y, equals the right-hand side of the equation.

> The gradient of the tangent and a point through which it passes is substituted into the equation for a straight line.

Test yourself

1 The points A, B, C, D have coordinates (1, 0), (4, 1), (−1, 3), (2, 4) respectively.
(a) Show that lines AB and CD are parallel.
(b) Find the equation of AB in the form $ax + by + c = 0$.

2 The points A and B have coordinates (−7, 4) and (k, −1) respectively.
(a) If the gradient of AB is $-\frac{1}{2}$, find the value of the constant k.
(b) The line BC is perpendicular to AB. Find the equation of line BC.

3 The points A, B, C have coordinates (−3, 2), (1, 6), (6, 1).
(a) Show that AB is perpendicular to BC.
(b) Find the length of AB and the length of BC.
(c) Find the value of tan $\widehat{ACB}$ in the form $\frac{a}{b}$

4 The points A, B, C, D have coordinates (−7, 4), (3, −1), (6, 1), (k, −15) respectively.
(a) Find the gradient of AB. [2]
(b) Find the equation of AB and simplify your answer. [3]
(c) Find the length of AB. [2]
(d) The point E is the mid-point of AB Find the coordinates of E. [2]
(e) Given that CD is perpendicular to AB, find the value of the constant k. [4]

5 The circle C has equation $x^2 + y^2 - 8x - 6y = 0$.
The straight line L has equation $y + 2x + 4 = 0$.
(a) Write down the coordinates of the centre of circle C and its radius. [3]
(b) Show that the straight line L and the circle C do not intersect or meet. [4]

6 A circle has the equation $x^2 + y^2 - 4x + 6y = 3$.
(a) Find the coordinates of the centre of the circle and its radius.
(b) Show that the point P(2, 1) lies on the circle.

> ### Active Learning
>
> There are lots of formulae and techniques to remember in this chapter. Write a summary of the formulae and techniques used in this chapter. Remember to also include circle properties and information about angles such as the base angles of an isosceles triangle being equal, angles in a semi-circle are 90°, etc. Try to fit it all on an A4 sheet of paper. Take a photo of the paper using your phone and refer to it when you have some free time.

7 Circle C has centre A(2, 3) and radius 5.
(a) Find the equation of circle C in the form $x^2 + y^2 + ax + by + c = 0$ where a, b and c are constants to be determined.
(b) Find the equation of the tangent to the circle at the point P(5, 7).

8 The circle C has centre A and equation $x^2 + y^2 - 8x + 2y + 7 = 0$.
(a) Find the coordinates of A and the radius of C.   [3]
(b) The point P has coordinates (7, −2).
  (i) Verify that P lies on C.
  (ii) Given that the point Q is such that PQ is a diameter of C, find the coordinates of Q.   [4]
(c) The line L has equation $y = 2x - 4$. Find the coordinates of the points of intersection of L and C.   [4]

9 The circle C has centre A and radius r. The points P (1, −4) and Q (9, 10) are at either end of a diameter of C.
(a) (i)   Write down the coordinates of A.
  (ii) Show that $r = \sqrt{65}$.
  (iii) Write down the equation of C.   [4]
(b) Verify that the point R (4, 11) lies on C.   [2]
(c) Find $Q\hat{P}R$.   [2]

Summary

Coordinate geometry and straight lines

The gradient of the line joining two points

The gradient of the line joining points (x_1, y_1) and (x_2, y_2) is given by:

$$\text{Gradient} = \frac{y_2 - y_1}{x_2 - x_1}$$

The length of a line joining two points

The length of a straight line joining the two points (x_1, y_1) and (x_2, y_2) is given by:

$$\sqrt{(x_2 - x_1)^2 + (y_2 - y_1)^2}$$

The mid-point of the line joining two points

The mid-point of a line joining the points (x_1, y_1) and (x_2, y_2) is given by:

$$\left(\frac{x_1 + x_2}{2}, \frac{y_1 + y_2}{2} \right)$$

The equation of a straight line

The equation of a straight line with gradient m and which passes through a point (x_1, y_1) is given by:

$$y - y_1 = m(x - x_1)$$

Condition for two straight lines to be parallel to each other

The lines must both have the same gradient.

Condition for two straight lines to be perpendicular to each other

If one line has a gradient of m_1 and the other a gradient of m_2 then the lines are perpendicular to each other if $m_1 m_2 = -1$.

The two forms for the equation of a circle

A circle having an equation in the form:

$$(x - a)^2 + (y - b)^2 = r^2 \qquad \text{has centre } (a, b) \text{ and radius } r.$$

A circle having an equation in the form:

$$x^2 + y^2 + 2gx + 2fy + c = 0 \qquad \text{has centre } (-g, -f) \text{ and radius } \sqrt{g^2 + f^2 - c}.$$

Circle properties

The angle in a semicircle is a right angle.

The perpendicular from the centre of a circle to a chord bisects the chord.

The radius to a point on the circle and the tangent through the same point are perpendicular to each other.

4 Sequences and series – the Binomial Theorem

Introduction

You are familiar with expanding $(a + b)^2$ or $(a + b)^3$, but what if you have to expand $(a + b)^6$? Luckily there is an easy way to do this expansion using a formula which you will learn how to use in this topic.

4.1 The binomial expansion

The binomial expansion is the expansion of an expression of the form $(a + b)^n$ where n is a positive integer. The formula for the expansion will be given in the formula booklet and is shown here:

$$(a + b)^n = a^n + \binom{n}{1}a^{n-1}b + \binom{n}{2}a^{n-2}b^2 + \ldots + \binom{n}{r}a^{n-r}b^r + \ldots + b^n$$

where,

$$\binom{n}{r} = {}^nC_r = \frac{n!}{r!(n-r)!}$$

You do not need to memorise these formulae as they are given in the formula booklet

$n!$ means n factorial. If $n = 5$ then $5! = 5 \times 4 \times 3 \times 2 \times 1$

Note that $0! = 1$

Example

1   To see how this formula is used, we will use an example. Expand $(a + b)^4$

. . . .

Answer

1   First carefully copy down the formula from the formula booklet:

$$(a + b)^n = a^n + \binom{n}{1}a^{n-1}b + \binom{n}{2}a^{n-2}b^2 + \ldots$$

You will also need this formula from the formula booklet.

$$\binom{n}{r} = {}^nC_r = \frac{n!}{r!(n-r)!}$$

Make sure you can find ${}^nC_r$ using a calculator

Substituting $n = 4$ into each formula gives:

$$(a + b)^4 = a^4 + \binom{4}{1}a^3 b + \binom{4}{2}a^2 b^2 + \binom{4}{3}a b^3 + \binom{4}{4}b^4$$

Now substituting the numbers $\binom{4}{1}$ for n and r into $\frac{n!}{r!(n-r)!}$ gives

$$\frac{4!}{1!(4-1)!} = \frac{4 \times 3 \times 2 \times 1}{3 \times 2 \times 1} = 4$$

This is repeated by substituting in numbers for $\binom{4}{2}$, $\binom{4}{3}$ and $\binom{4}{4}$ giving the numbers 6, 4 and 1 respectively.

Hence $(a + b)^4 = a^4 + 4a^3b + 6a^2b^2 + 4ab^3 + b^4$

4.2 Pascal's triangle

You can also find the coefficients in the expansion of $(a + b)^n$ by using Pascal's triangle.

Suppose you want to expand the expression from the previous example, $(a + b)^4$ using Pascal's triangle. You would write down Pascal's triangle and look for the line starting 1 and then 4 (because n is 4 here). The line 1, 4, 6, 4, 1 gives the

4 Sequences and series – the Binomial Theorem

coefficients. This avoids a calculation involving the factorials for each coefficient, but you will have to remember how to construct Pascal's triangle.

$$
\begin{array}{ccccccccccc}
 & & & & & 1 & & & & & \\
 & & & & 1 & & 1 & & & & \\
 & & & 1 & & 2 & & 1 & & & \\
 & & 1 & & 3 & & 3 & & 1 & & \\
 & 1 & & 4 & & 6 & & 4 & & 1 & \\
1 & & 5 & & 10 & & 10 & & 5 & & 1
\end{array}
$$

> Notice that all the rows start and end with a 1. Notice also that the other numbers are found by adding the pairs of numbers immediately above. For example, if we have 1 3 in the line above, then the number to be entered between these numbers on the next line is a 4.

Example

1 Use the binomial expansion to expand $(2 + 3x)^3$.

Answer

1 First obtain the formula for the binomial expansion from the formula booklet.

$$(a + b)^n = a^n + \binom{n}{1}a^{n-1}b + \binom{n}{2}a^{n-2}b^2 + \ldots$$

Here $a = 2$, $b = 3x$ and $n = 3$.

Substituting these values into the formula gives:

$$(2 + 3x)^3 = 2^3 + \binom{3}{1}2^2(3x) + \binom{3}{2}2^1(3x)^2 + \binom{3}{3}2^0(3x)^3$$

As $n = 3$ here, we look for the line in Pascal's triangle which starts at 1 and then 3, etc. You can see that the numbers in this line are: 1 3 3 1.

These are the values of $\binom{3}{0}$, $\binom{3}{1}$, $\binom{3}{2}$ and $\binom{3}{3}$. So for example $\binom{3}{1} = 3$ and $\binom{3}{3} = 1$.

Hence we can write the expansion like this:

$$(2 + 3x)^3 = (1)2^3 + (3)2^2(3x) + (3)2^1(3x)^2 + (1)2^0(3x)^3$$

Hence $(2 + 3x)^3 = 8 + 36x + 54x^2 + 27x^3$ | Remember that $2^0 = 1$ |

4.3 The binomial expansion where $a = 1$

When the first term in the bracket (i.e. a) is 1, the binomial expansion becomes:

$$(1 + x)^n = 1 + nx + \frac{n(n-1)}{2!}x^2 + \frac{n(n-1)(n-2)}{3!}x^3 + \ldots$$

Again this formula is given in the formula booklet so you don't need to memorise it.

Examples

1 (a) Write down the expansion of $(1 + x)^6$ in ascending powers of x up to and including the term in x^3. [2]

 (b) By substituting an appropriate value for x in your expansion in (a), find an approximate value for 0.99^6. Show all your working and give your answer correct to four decimal places. [3]

Answer

1 (a) The formula for the expansion of $(1 + x)^n$ is obtained from the formula booklet.

$$(1 + x)^n = 1 + nx + \frac{n(n-1)}{2!}x^2 + \frac{n(n-1)(n-2)}{3!}x^3 + \dots$$

Substituting $n = 6$ into this formula gives:

$$(1 + x)^6 = 1 + 6x + \frac{6(5)}{2!}x^2 + \frac{6(5)(4)}{3!}x^3 + \dots$$

Note that using the first three terms only provides an approximate value.

Hence, $(1 + x)^6 \approx 1 + 6x + \frac{6(5)x^2}{2!} + \frac{6(5)(4)x^3}{3!}$

$$\approx 1 + 6x + 15x^2 + 20x^3$$

(b) $1 - 0.01 = 0.99$

So, $0.99^6 = (1 - 0.01)^6$

Substituting $x = -0.01$ into the expansion of $(1 + x)^6$ gives

$$(1 - 0.01)^6 \approx 1 + 6(-0.01) + 15(-0.01)^2 + 20(-0.01)^3$$

$$\approx 0.94148$$

$$\approx 0.9415 \text{ (4 decimal places)}$$

BOOST

Grade ⇧⇧⇧⇧

When obtaining a numerical answer, always check to see if the question asks for the answer to be given to a certain number of decimal places or significant figures. Marks can be needlessly lost by not doing this.

---

2 (a) Expand $\left(x + \dfrac{2}{x}\right)^4$, simplifying each term of the expansion.   [4]

(b) The coefficient of x^2 in the expansion of $(1 + x)^n$ is 55. Given that n is a positive integer, find the value of n.   [3]

Answer

2 (a) Obtaining the formula and following the pattern in the terms gives:

$$(a + b)^n = a^n + \binom{n}{1}a^{n-1}b + \binom{n}{2}a^{n-2}b^2 + \binom{n}{3}a^{n-3}b^3 + \dots$$

$$(a + b)^4 = a^4 + \binom{4}{1}a^3b + \binom{4}{2}a^2b^2 + \binom{4}{3}ab^3 + \binom{4}{4}b^4$$

$$(a + b)^4 = a^4 + 4a^3b + 6a^2b^2 + 4ab^3 + b^4$$

Substituting $a = x$ and $b = \dfrac{2}{x}$ into the equation gives:

$$\left(x + \frac{2}{x}\right)^4 = x^4 + 4x^3\left(\frac{2}{x}\right) + 6x^2\left(\frac{2}{x}\right)^2 + 4x\left(\frac{2}{x}\right)^3 + \left(\frac{2}{x}\right)^4$$

$$= x^4 + 8x^2 + 24 + \left(\frac{32}{x^2}\right) + \left(\frac{16}{x^4}\right)$$

(b) In the expansion of $(1 + x)^n$ the coefficient of x^2 is $\dfrac{n(n-1)}{2}$

Hence $\dfrac{n(n-1)}{2} = 55$

$$n^2 - n = 110$$

BOOST

Grade ⇧⇧⇧⇧

You can find these numbers using Pascal's triangle but you will need to know how to construct and use it, as it is not given in the formula booklet.

With practice the values of the coefficients will become known or can be calculated quickly.

$$n^2 - n - 110 = 0$$

Factorising this quadratic equation gives $(n - 11)(n + 10) = 0$

Solving gives $n = 11$ or $n = -10$

The question says n is a positive integer, so $n = 11$.

> Notice that this is a quadratic equation so it needs to be rearranged to equal zero so it can be factorised and solved.

3   Write down and simplify the first four terms in the binomial expansion of

$$\left(1 + \frac{x}{2}\right)^6$$

Answer

3

$$(1 + x)^n = 1 + nx + \frac{n(n-1)}{2!}x^2 + \frac{n(n-1)(n-2)}{3!}x^3$$

> Substitute n as 6 and x as $\left(\frac{x}{2}\right)$ into the formula.

$$\left(1 + \frac{x}{2}\right)^6 = 1 + 6\left(\frac{x}{2}\right) + \frac{(6)(5)}{2 \times 1}\left(\frac{x}{2}\right)^2 + \frac{(6)(5)(4)}{3 \times 2 \times 1}\left(\frac{x}{2}\right)^3$$

$$= 1 + 3x + \frac{15}{4}x^2 + \frac{5}{2}x^3$$

4   In the binomial expansion of $(a + 2x)^5$, the coefficient of the term in x^2 is four times the coefficient of the term in x. Find the value of the constant .

BOOST
Grade ⇧⇧⇧⇧

> You could have used Pascal's triangle here but this is not in the formula booklet so you would need to know how to construct it.

> The formula:
> $$\binom{n}{r} = \frac{n!}{r!(n-r)!}$$
> has been used here. This formula is obtained from the formula booklet.

Answer

4

$$(a + b)^n = a^n + \binom{n}{1}a^{n-1}b + \binom{n}{2}a^{n-2}b^2 + \dots$$

Here $a = a$, $b = 2x$ and $n = 5$

Substituting these values into the formula gives:

$$(a + 2x)^5 = a^5 + \binom{5}{1}a^4(2x) + \binom{5}{2}a^3(2x)^2 + \dots$$

Now $\binom{5}{1} = \frac{5!}{1!(5-1)!} = \frac{5!}{1!4!} = 5$

$\binom{5}{2} = \frac{5!}{2!(5-2)!} = \frac{5!}{2!3!} = 10$

Hence: $(a + 2x)^5 = a^5 + (5)a^4(2x) + (10)a^3(2x)^2 + \dots$

$$= a^5 + 10a^4x + 40a^3x^2 + \dots$$

Coefficient of x^2 is four times the coefficient of x, so

$$40a^3 = 4 \times 10a^4$$

$$40a^3 = 40a^4$$

Dividing both sides by $40a^3$ gives $a = 1$. $(a \neq 0)$

5   Expand $(a + b)^4$. Hence expand $\left(2x + \frac{1}{2x}\right)^4$, simplifying each term of the expansion.

[4]

. . . .

Answer

5

$$(a + b)^n = a^n + \binom{n}{1}a^{n-1}b + \binom{n}{2}a^{n-2}b^2 + \binom{n}{3}a^{n-3}b^3 + \ldots$$

$$(a + b)^4 = a^4 + \binom{4}{1}a^3b + \binom{4}{2}a^2b^2 + \binom{4}{3}ab^3 + \binom{4}{4}b^4$$

Finding $\binom{4}{1}$, $\binom{4}{2}$, $\binom{4}{3}$, $\binom{4}{4}$, by using the formula or by using Pascal's triangle and substituting them in to the above formula gives:

$$(a + b)^4 = a^4 + 4a^3b + 6a^2b^2 + 4ab^3 + b^4$$

$$\left(2x + \frac{1}{2x}\right)^4 = (2x)^4 + 4(2x)^3\left(\frac{1}{2x}\right) + 6(2x)^2\left(\frac{1}{2x}\right)^2 + 4(2x)\left(\frac{1}{2x}\right)^3 + \left(\frac{1}{2x}\right)^4$$

$$= 16x^4 + 16x^2 + 6 + \frac{1}{x^2} + \frac{1}{16x^4}$$

The formula sheet contains the following formulae related to the binomial expansion.

$$(a + b)^n = a^n + \binom{n}{1}a^{n-1}b + \binom{n}{2}a^{n-2}b^2 + \ldots + \binom{n}{r}a^{n-r}b^r + \ldots + b^n \qquad (n \in \mathbb{N})$$

where $\binom{n}{r} = {}^nC_r = \dfrac{n!}{r!(n-r)!}$

$$(1 + x)^n = 1 + nx + \frac{n(n-1)}{1 \times 2}x^2 + \ldots + \frac{n(n-1)\ldots(n-r+1)}{1 \times 2 \times \ldots r}x^r + \ldots \qquad (|x| < 1, n \in \mathbb{R})$$

Evaluate which of the two formulae you would prefer to use to expand $(3 + 2x)^6$. Use both formulae in turn and think about which one you found easier. Think about which method is less prone to errors.

Active Learning

Test yourself

① In the binomial expansion of $(2 + 3x)^5$, find the coefficient of the term in x^2.

② Write down and simplify the first four terms in the binomial expansion of $(1 + 3x)^6$.

③ Use the binomial theorem to expand $(3 + 2x)^3$, simplifying each term of your expansion. [3]

④ Use the binomial expansion to express $(\sqrt{5} - \sqrt{2})^5$ in the form $a\sqrt{5} + b\sqrt{2}$ where a, b are integers to be found.

⑤ (a) Use the binomial theorem to express $(1 + \sqrt{6})^5$ in the form $a + b\sqrt{6}$, where a, b are integers whose values are to be found.
   (b) The coefficient of x^2 in the expansion of $(1 + 3x)^n$ is 495. Given that n is a positive integer, find the value of n.

BOOST

Grade ⬆⬆⬆⬆

You have to use the binomial theorem here as it is specified in the question. If you found the answer by multiplying out the brackets you would not gain any marks.

Summary

Binomial expansions

The binomial expansion of $(a + b)^n$ for positive integer n

$$(a + b)^n = a^n + \binom{n}{1}a^{n-1}b + \binom{n}{2}a^{n-2}b^2 + \ldots + \binom{n}{r}a^{n-r}b^r + \ldots + b^n$$

$$\binom{n}{r} = {}^nC_r = \frac{n!}{r!(n-r)!}$$

The binomial expansion of $(1 + x)^n$ for positive integer n

$$(1 + x)^n = 1 + nx + \frac{n(n-1)}{2!}x^2 + \frac{n(n-1)(n-2)}{3!}x^3 + \ldots + \frac{n(n-1)\ldots(n-r+1)}{r!}x^r + \ldots$$

5 Trigonometry

Introduction

This topic builds on and reinforces your GCSE work on trigonometry.

5.1 Sine, cosine and tangent functions and their exact values

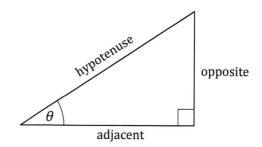

The following ratios, covered in your GCSE studies, only apply to right-angled triangles:

$$\sin \theta = \frac{\text{opposite}}{\text{hypotenuse}} \qquad \cos \theta = \frac{\text{adjacent}}{\text{hypotenuse}} \qquad \tan \theta = \frac{\text{opposite}}{\text{adjacent}}$$

The exact values of the sine, cosine and tangent of 30°, 45° and 60°

The exact values of the above angles can be determined by drawing triangles, working out the lengths of the sides that aren't known and then using trigonometry to work out the exact values of the angles.

The exact values of the sine, cosine and tangent of 45°

The exact values can be worked out by drawing the following triangle:

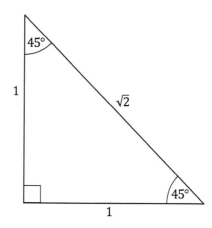

The length of the hypotenuse is worked out using Pythagoras' theorem.
Hypotenuse = $\sqrt{1^2 + 1^2} = \sqrt{2}$

If you are asked to find the exact value you must not approximate any of the sides as a decimal.

$$\sin 45° = \frac{\text{opposite}}{\text{hypotenuse}} = \frac{1}{\sqrt{2}}$$

$$\cos 45° = \frac{\text{adjacent}}{\text{hypotenuse}} = \frac{1}{\sqrt{2}}$$

$$\tan 45° = \frac{\text{opposite}}{\text{adjacent}} = \frac{1}{1} = 1$$

The exact values of the sine, cosine and tangent of 30° and 60°

The exact values can be worked out using an equilateral triangle having sides of length 2 and then drawing in one of the lines of symmetry to form two identical right-angled triangles:

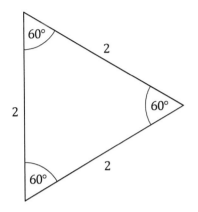

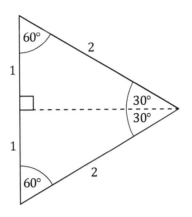

BOOST

Grade ⇧⇧⇧⇧

Familiarise yourself with these diagrams and get used to working out any lengths missing from the right-angled triangles formed.

Start off with an equilateral triangle having length of side = 2. All the angles of the triangle will be 60°.

The perpendicular bisector divides the original triangle into two right-angled triangles. Notice that the angle is bisected as well as a side.

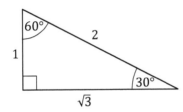

Half of the original triangle is used. The length of the base of this triangle is worked out using Pythagoras' theorem.
Base = $\sqrt{2^2 - 1^2} = \sqrt{3}$

$$\sin 30° = \frac{1}{2} \qquad \sin 60° = \frac{\sqrt{3}}{2}$$

$$\cos 30° = \frac{\sqrt{3}}{2} \qquad \cos 60° = \frac{1}{2}$$

$$\tan 30° = \frac{1}{\sqrt{3}} \qquad \tan 60° = \sqrt{3}$$

If you are asked to find the exact value you must not approximate any of the sides as a decimal.

5.2 Obtaining angles given a trigonometric ratio

Finding angles using the CAST method

One method of obtaining angles given a trigonometric ratio is called the CAST method.

The CAST method uses the diagram shown below. A indicates that all the ratios are positive in the first quadrant (for angles between 0° and 90°), S indicates that sine is positive in the second quadrant (90° to 180°), T indicates that tangent is positive in the third quadrant (180° to 270°) and C indicates that cosine is positive in the fourth quadrant (270° to 360°).

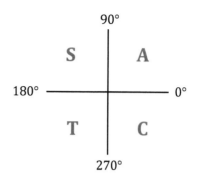

> This simple diagram should be remembered if you intend to use the CAST method.

CAST stands for **C**os, **A**ll, **S**in, **T**an and the diagram shows where these functions are positive. For example, suppose we wanted all the values of the angle θ in the range $0° \leq \theta \leq 360°$ where $\sin \theta = 0.6946$. Here we have a positive value for $\sin \theta$. The sine function is positive in the first and second quadrants.

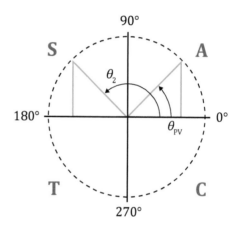

> Notice that the angles are measured anticlockwise from 0°.

Two triangles are drawn in the regions where $\sin \theta$ is positive. You can use your calculator to find the first value, θ_{PV}, by working out $\sin^{-1}(0.6946)$ using a calculator. This gives $\theta_{PV} = 44°$. As both triangles are identical the value of θ_2 is found by subtracting 44° from 180°. Hence the other angle is $180° - 44° = 136°$.

Therefore $\theta = 44°$ or $136°$.

Example

1   Find all the values of the angle θ in the range $0° \leq \theta \leq 360°$ where $\cos \theta = -\dfrac{1}{2}$.

....

Answer

1   Two triangles are drawn in the regions where $\cos \theta$ is negative. Cosine is negative in the second and third quadrants. θ_{PV} can be found by using a calculator and entering $\cos^{-1}(-0.5)$ giving a value of 120°. By symmetry, the value of θ_2 can be found by subtracting 120° from 360°. Hence the other angle is $360° - 120° = 240°$.

> Using this method, the solutions are found by making the same angle from the horizontal in each of the appropriate quadrants, e.g. $180° - 60° = 120°$ and $180° + 60° = 240°$.

Therefore $\theta = 120°$ or $240°$.

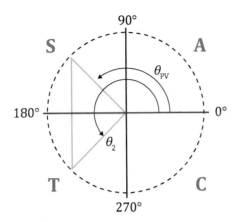

Finding angles using trigonometric graphs

Another method involves using the trigonometric graphs to find all the angles. Here you must be able to draw the graphs of each trigonometric function (sin, cos and tan).

You will probably be familiar with these graphs from your GCSE work. The graphs are included on pages 103 to 105.

Example

1 Find all the values of the angle θ in the range $0° \leq \theta \leq 360°$ where $\sin\theta = \frac{1}{2}$.

Answer

1

The graph of $y = \sin\theta$ is drawn in the range $0° \leq \theta \leq 360°$.

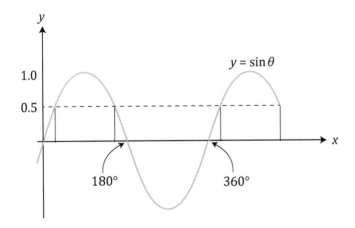

The first angle is found by performing the calculation $\sin^{-1}\left(\frac{1}{2}\right)$ or $\sin^{-1}(0.5)$ using a calculator or by recognition (see page 95).

The result is 30°. By the symmetry of the graph you can see that the other angle will be 180° − 30° = 150°. Hence the two values of θ in the required range are 30° and 150°.

BOOST

Grade ⇧⇧⇧⇧

It is essential to be able to draw the graphs of sin, cos and tan and mark their points of intersection with the axes.

5.3 The sine and cosine rules

The sine and cosine rules can be used with any triangle, not just those containing a right angle.

The angles are denoted by the letters A, B and C and the lengths of the sides opposite these angles are denoted by a, b and c respectively.

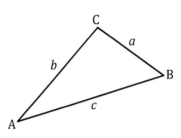

The sine rule should be used in the most useful form for the given question:

$$\frac{a}{\sin A} = \frac{b}{\sin B} = \frac{c}{\sin C}$$

or

$$\frac{\sin A}{a} = \frac{\sin B}{b} = \frac{\sin C}{c}$$

The formulae for the sine and cosine rules are not included in the formula booklet so must be memorised.

The sine rule states:

$$\frac{a}{\sin A} = \frac{b}{\sin B} = \frac{c}{\sin C}$$

The cosine rule states:

$$a^2 = b^2 + c^2 - 2bc \cos A$$

Example

1   The diagram below shows a sketch of the triangle ABC with AB = 8 cm, AC = x cm, BC = $(x + 2)$ cm and $A\widehat{B}C = 60°$.

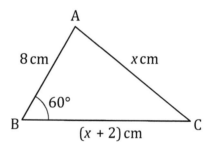

(a) Write down and simplify an equation satisfied by x. Hence evaluate x.   [3]

(b) Find the size of $A\widehat{C}B$.   [2]

· ·

Answer

1   (a) Using the cosine rule:

$$x^2 = 8^2 + (x + 2)^2 - 2 \times 8 \times (x + 2) \cos 60°$$

$$x^2 = 64 + x^2 + 4x + 4 - 16 \times (x + 2) \times \frac{1}{2}$$

$$x^2 = 64 + x^2 + 4x + 4 - 8x - 16$$

$$x^2 = x^2 - 4x + 52$$

$$0 = -4x + 52$$

$$x = 13$$

(b) Using the sine rule:

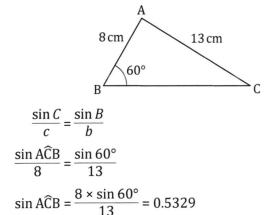

$$\frac{\sin C}{c} = \frac{\sin B}{b}$$

$$\frac{\sin A\widehat{C}B}{8} = \frac{\sin 60°}{13}$$

$$\sin A\widehat{C}B = \frac{8 \times \sin 60°}{13} = 0.5329$$

Hence $A\widehat{C}B = 32.2°$

Step by STEP

The diagram below shows a sketch of the triangle ABC with $\sin A = \frac{4}{5}$, $\sin B = \frac{8}{17}$, $\cos C = -\frac{13}{85}$, AC = x cm and BC = y cm.

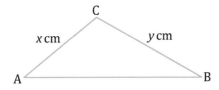

Given that AB = 10.5 cm, find the exact value of x.

Steps to take

1 Lots of information is given in the question so read the question a couple of times to take it all in.

2 It is worth re-drawing the diagram marking any sides or angles that are in the question but not in the diagram.

3 Notice that the sines of angles A and B are known and also the sides x and y marked. This means if the sine rule is used, it will be possible to get y in terms of x.

4 Two sides are known in terms of x and the length of the third side is known. As $\cos C$ is known we can use the cosine rule and form an equation in x which can be solved.

Answer

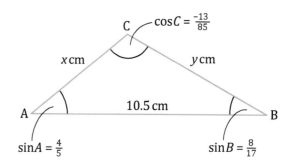

Using the sine rule $\dfrac{a}{\sin A} = \dfrac{b}{\sin B}$

$$\dfrac{y}{\frac{4}{5}} = \dfrac{x}{\frac{8}{17}}$$

$$\dfrac{5y}{4} = \dfrac{17x}{8}$$

$$y = 1.7x$$

Using the cosine rule: $c^2 = a^2 + b^2 - 2ab \cos C$

$$10.5^2 = (1.7x)^2 + x^2 - 2(1.7x)(x) \cos C$$

$$110.25 = 2.89x^2 + x^2 - 3.4x^2\left(-\dfrac{13}{85}\right)$$

$$110.25 = 3.89x^2 + 0.52x^2$$

$$110.25 = 4.41x^2$$

$$x = 5 \text{ cm (the negative solution is ignored, as } x \text{ is a length)}$$

5.4 The area of a triangle

If two sides of a triangle are known as well as the included angle, then the area of the triangle can be found using the formula:

> Area of triangle $= \dfrac{1}{2}ab \sin C$

This formula works for all triangles but for areas of right-angled triangles, use the formula
$$A = \dfrac{1}{2} \times base \times height$$
Note that you will not be given this formula.

Warning: be careful when using this formula to work out the angle when the area and two sides of the triangle are known. For example $\sin C = \dfrac{1}{2}$ can have two solutions 30° and 150°. If another angle in the triangle is known then the obtuse angle may not be a possible solution. Be guided by the wording of the question, so look for the plural 'angles' in the question to see if you are looking for two possible angles.

Examples

1 The diagram below shows the triangle ABC with AB = x cm, AC = $(x + 4)$ cm and $\hat{BAC} = 150°$.

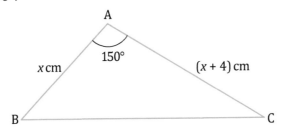

Given that the area of the triangle ABC is 15 cm²,

(a) find the value of x. [3]

(b) find the length of BC correct to one decimal place. [2]

When you look at the diagram, you can see that the two sides and the included angle (i.e. the angle between the two sides) are marked. The area is known, so we use the formula for finding the area of a triangle that is not right-angled.

. . . .

Answer

1 (a) Area of triangle ABC = $\frac{1}{2}bc \sin A$

$$= \frac{1}{2}(x + 4)x \sin 150°$$

$$= \frac{1}{2}\frac{(x + 4)x}{2}$$

$$= \frac{1}{4}(x^2 + 4x)$$

$$15 = \frac{1}{4}(x^2 + 4x)$$

$$60 = x^2 + 4x$$

$$x^2 + 4x - 60 = 0$$

$$(x - 6)(x + 10) = 0$$

Solving gives $x = 6$

As x is the length of one of the sides, the −10 solution is impossible so it is discarded.

(b) Substituting the value $x = 6$ for the two sides gives

AB = 6 cm and AC = 10 cm

Using the cosine rule:

$$BC^2 = 6^2 + 10^2 - 2 \times 6 \times 10 \cos 150°$$

$$BC^2 = 36 + 100 - 120 \cos 150°$$

$$BC^2 = 136 + 103.92$$

$$BC = \sqrt{239.92}$$

$$BC = 15.5 \text{ cm (correct to 1 decimal place)}$$

2   The triangle ABC is such that AB = 16 cm, AC = 9 cm and $\widehat{ABC}$ = 23°.

(a)   Find the possible values of $\widehat{ACB}$. Give your answers correct to the nearest degree.   [2]

(b)   Given that $\widehat{BAC}$ is an **acute** angle, find:

(i)   the size of $\widehat{BAC}$, giving your answer correct to the nearest degree,

(ii)   the area of triangle ABC, giving your answer correct to one decimal place.   [4]

It is a good idea to draw a sketch showing what the triangle might look like. It will enable you to see which sides and angles are known so you can decide whether to use the sine or cosine rule.

Answer

2   (a)

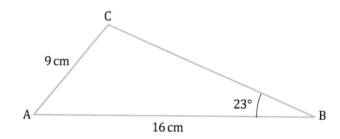

Using the sine rule:

$$\frac{\sin \widehat{ACB}}{16} = \frac{\sin 23°}{9}$$

$$\sin \widehat{ACB} = \frac{16 \sin 23°}{9}$$

$$\sin \widehat{ACB} = 0.6946$$

$$\widehat{ACB} = 44° \text{ or } 136°$$

The question does not give you a diagram for the triangle and notice that it asks for possible **values** of $\widehat{ACB}$. This implies more than one value, so you have to look for the two angles that have a sine of 0.6946.

(b)   (i)   If angle $\widehat{BAC}$ is acute, angle $\widehat{ACB}$ must be 136°

Hence, angle $\widehat{BAC}$ = 180° − (136° + 23°) = 21°

(ii)   Area of triangle = $\frac{1}{2} bc \sin A$

$$= \frac{1}{2} \times 9 \times 16 \times \sin 21°$$

$$= 25.8025 \text{ cm}^2$$

$$= 25.8 \text{ cm}^2 \text{ (to one decimal place)}$$

5.5 Sine, cosine and tangent: their graph, symmetries and periodicity

Radian measure

There is another unit for measuring angles called the radian.

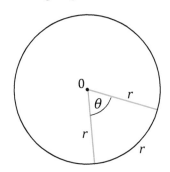

When the length of an arc is the same as the radius then the angle between the two radii and the arc, θ, is 1 radian. An arc of length $2r$ would give an angle at the centre of 2 radians and an arc of length θr would give an angle at the centre of θ radians.

If the length of the arc equals half the circumference then the length of the arc is πr. If this length of arc corresponds to an angle at the centre of θ radians then equating the two gives:

$$r\theta = \pi r$$

So $\theta = \pi$ (now as $\theta = 180°$) we can write π radians = 180°

So 1 radian = $\dfrac{180}{\pi} = \dfrac{180}{3.14} = 57.3°$

Here are some popular angles expressed in radians and degrees:

2π radians = 360°

$\dfrac{\pi}{2}$ radians = 90° $\dfrac{\pi}{4}$ radians = 45°

$\dfrac{\pi}{3}$ radians = 60° $\dfrac{\pi}{6}$ radians = 30°

Check that your calculator is set to radian mode when performing radian calculations. Remember to turn it back to degrees after completing the question.

The sine graph ($y = \sin \theta$) where θ is expressed in radians

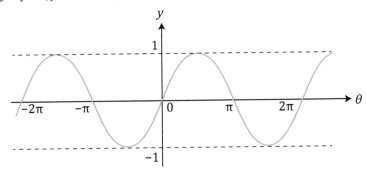

The sine graph has a period of 2π meaning the graph repeats itself every 2π radians.

The sine graph has a period of 2π as a particular value of θ will have the same y-value at an angle of $\theta + 2\pi$, $\theta + 4\pi$, and so on.

The sine graph ($y = \sin \theta$) where θ is expressed in degrees

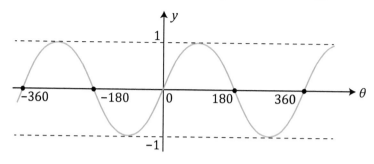

The cosine graph ($y = \cos \theta$) where θ is expressed in radians

The cosine graph has a period of 2π meaning the graph repeats itself every 2π radians.

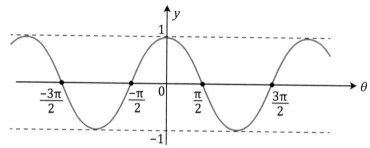

The cosine graph has a period of 2π as a particular value of θ will have the same y-value at an angle of $\theta + 2\pi$, $\theta + 4\pi$, and so on.

The cosine graph ($y = \cos \theta$) where θ is expressed in degrees

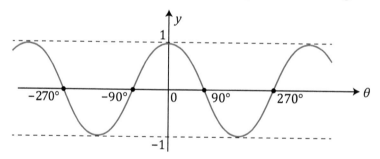

The tangent graph ($y = \tan \theta$) where θ is expressed in radians

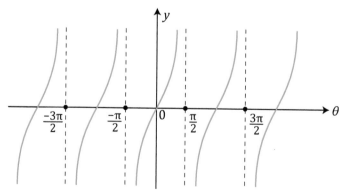

The period of the tan θ graph is π radians.

The tangent graph ($y = \tan \theta$) where θ is expressed in degrees

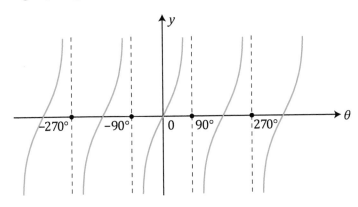

5.6 Using $\tan \theta = \dfrac{\sin \theta}{\cos \theta}$ and $\sin^2 \theta + \cos^2 \theta = 1$

There are two trigonometric identities that you may need to use when solving simple trigonometric equations.

These two identities are:

$$\tan \theta = \frac{\sin \theta}{\cos \theta}$$

$$\cos^2 \theta + \sin^2 \theta = 1$$

> Both of these identities must be remembered. They are not included in the formula booklet.

You will see both of the above identities being used in the examples following the next section.

5.7 Solving trigonometric equations

> You may see the identity symbol ≡ being used sometimes instead of the equals sign. You can treat them as being the same. An identity is a relationship involving a letter which is true for all values of the letter.

The graphs of trigonometric functions can be used to help identify all the solutions of a simple trigonometric equation in a given interval.

Suppose we have to solve the following equation in the interval $0° \leq \theta \leq 360°$

$$\sin (2x - 30)° = \frac{1}{2}$$

Draw a graph of $y = \sin \theta$. You will need to go a lot further than 360° for your graph so that all the possible solutions are shown.

Here we will go as far as 720°

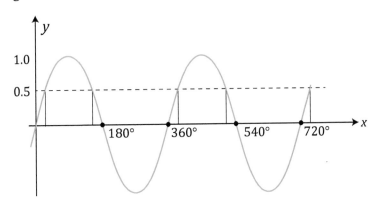

You could also use the CAST method for finding all the angles θ in the required range.

$$2x - 30 = \sin^{-1}\left(\frac{1}{2}\right)$$

Letting $\quad \theta = 2x - 30$

So $\quad\quad \theta = \sin^{-1}\left(\frac{1}{2}\right)$

Using the calculator (or the triangles for the exact values learnt earlier) we have

$\theta = 30°$ we can then see from the graph that the line $y = \frac{1}{2}$ also intersects the curve at the following values of θ:

$\theta = 150°, 390°, 510°$

When looking for all the solutions, between $0°$ and $360°$, of $\sin(2\theta - 30) = \frac{1}{2}$, the values of θ that need to be considered will lie between $2 \times 0 - 30$ and $2 \times 360 - 30$, i.e. $-30°$ and $690°$.

Sine is positive in the first and second quadrants, therefore $\theta = 30°$ or $\theta = 180° - 30 = 150°$, or (adding $360°$) $390°$ or $510°$, etc.

In practice, it's probably best to consider twice the range, since the multiple of the angle involved is 2, i.e. from $0°$ to $720°$. It is quicker to do this calculation.

Hence $\quad \theta = 30°, 150°, 390°, 510°$

So $\quad 2x - 30 = 30°, 150°, 390°, 510°$

$\quad\quad\quad\quad 2x = 60°, 180°, 420°, 540°$

$\quad\quad\quad\quad\quad x = 30°, 90°, 210°, 270°$

So values of x in the required range are $x = 30°, 90°, 210°, 270°$.

Examples

1 Find the values of x in the range $0° \le x \le 360°$, that satisfy the equation

$\quad 2\sin x = \tan x$

. . . .

Answer

1 $\quad\quad\quad\quad 2\sin x = \tan x$

Do not be tempted here to divide both sides by $\sin x$. If you do this then you will lose some solutions of the equation.

$\quad\quad\quad\quad 2\sin x = \dfrac{\sin x}{\cos x}$

$\quad\quad\quad\quad 2\sin x \cos x = \sin x$

$\quad\quad\quad\quad \sin x(2\cos x - 1) = 0$

Hence, $\sin x = 0 \quad$ or $\quad \cos x = \dfrac{1}{2}$

$\sin x = 0$ at $x = 0°, 180°, 360°$

Using the CAST method e.g. for $\cos x = \frac{1}{2}$, cosine is positive in the first and fourth quadrants, so $x = 60°$ or $x = 360° - 60° = 300°$.

$\cos x = \dfrac{1}{2}$ at $x = 60°$ or $300°$

Hence $x = 0°, 60°, 180°, 300°$ or $300°$

---

2 (a) Find all values of θ in the range $0° \le \theta \le 360°$ satisfying
$\quad 12\cos^2\theta - 5\sin\theta = 10$ [6]

(b) Find all values of x in the range $0° \le x \le 180°$ satisfying $\tan 2x = -1.6$ [2]

(c) Find all values of ϕ in the range $0° \le \phi \le 360°$ satisfying $\tan\phi + 2\sin\phi = 0$ [4]

Answer

2   (a)
$$12 \cos^2 \theta - 5 \sin \theta = 10$$
$$12(1 - \sin^2 \theta) - 5 \sin \theta = 10$$
$$12 \sin^2 \theta + 5 \sin \theta - 2 = 0$$
$$(4 \sin \theta - 1)(3 \sin \theta + 2) = 0$$

so
$$\sin \theta = \frac{1}{4} \quad \text{or} \quad \sin \theta = -\frac{2}{3}$$

When $\sin \theta = \frac{1}{4}$, $\theta = 14.5°$ or $165.5°$

When $\sin \theta = -\frac{2}{3}$, $\theta = 221.8°$ or $318.2°$

$$\theta = 14.5°, 165.5°, 221.8° \text{ or } 318.2°$$

> $\cos^2 \theta + \sin^2 \theta = 1$
> so $\cos^2 \theta = 1 - \sin^2 \theta$

> Remember to include all the solutions in the required range.

(b) $\tan 2x = -1.6$
$$2x = 122°, 302°$$
$$x = 61° \text{ or } 151°$$

(c) $\tan \phi + 2 \sin \phi = 0$
$$\frac{\sin \phi}{\cos \phi} + 2 \sin \phi = 0$$
$$\sin \phi + 2 \sin \phi \cos \phi = 0$$
$$\sin \phi (1 + 2 \cos \phi) = 0$$
$$\sin \phi = 0 \quad \text{or} \quad \cos \phi = -\frac{1}{2}$$
$$\phi = 0°, 120°, 180°, 240° \text{ and } 360°$$

> Use $\tan \phi = \dfrac{\sin \phi}{\cos \phi}$

Step by STEP

Find all values of θ between 0° and 360° satisfying: $7 \sin^2 \theta + 1 = 3 \cos^2 \theta - \sin \theta$

Steps to take

1   Notice the term in $\sin \theta$, it is easy to switch between $\sin^2 \theta$ and $\cos^2 \theta$ using $\sin^2 \theta + \cos^2 \theta = 1$, but it is hard to turn $\sin \theta$ into $\cos \theta$. So we need to turn the equation into one in $\sin \theta$.

2   We will probably obtain a quadratic equation in $\sin \theta$ which could be factorised and solved. Note that there is a possibility that the resulting quadratic may not be factorised, in which case the formula could be used.

3   Once the two values of $\sin \theta$ are found the values of θ in the correct range can be found using the symmetry of the graph or using the CAST method.

Answer

$$7 \sin^2 \theta + 1 = 3 \cos^2 \theta - \sin \theta$$
$$7 \sin^2 \theta + 1 = 3(1 - \sin^2 \theta) - \sin \theta$$

$$7 \sin^2 \theta + 1 = 3 - 3 \sin^2 \theta - \sin \theta$$

$$10 \sin^2 \theta + \sin \theta - 2 = 0$$

$$(2 \sin \theta + 1)(5 \sin \theta - 2) = 0$$

$$\sin \theta = -\tfrac{1}{2}, \sin \theta = \tfrac{2}{5}$$

Using symmetry of graphs or the CAST method

$$\sin \theta = -\tfrac{1}{2}, \theta = 210°, 330°$$

$$\sin \theta = \tfrac{2}{5}, \theta = 23.57°, 156.42°$$

Step by STEP

The diagram below shows a sketch of the triangle ABC with AB = 10 cm, AC = x cm, BC = $(x + 4)$ cm and $\widehat{ABC} = \alpha$, where $\cos \alpha = \tfrac{3}{5}$.

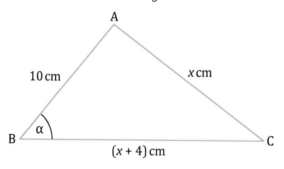

Find the exact value of the area of triangle ABC.

> Notice the word *exact* in the question. This means any surds should be left as surds and fractions should **not** be worked out as decimals.

Steps to take

1   Look at the question carefully and analyse what you know and what you think you need to find.

2   Here only one side and the cos of an angle are known. If we can find the value of x then we could use the formula Area = $\tfrac{1}{2} ab \sin C$ to find the area.

3   Use the cosine rule to form an equation in x which can then be solved.

4   Substitute the value of x to find the lengths of the two sides that include the angle.

5   Find $\sin \alpha$ using a right-angled triangle. The phrase 'exact value' in the question gives a clue that this method needs to be used.

6   Use the formula Area = $\tfrac{1}{2} ab \sin C$ to find the area.

. . . .

Answer

Using the cosine rule, we have $x^2 = 10^2 + (x + 4)^2 - 2 \times 10 \times (x + 4) \cos \alpha$

$$x^2 = 100 + x^2 + 8x + 16 - 2 \times 10 \times (x + 4) \cos \alpha$$

$$20(x + 4) \cos \alpha = 8x + 116$$

But $\cos \alpha = \tfrac{3}{5}$, hence $20(x + 4)\tfrac{3}{5} = 8x + 116$

$$12x + 48 = 8x + 116$$

$x = 17$ so the $(x + 4)$ side becomes $17 + 4 = 21$ cm

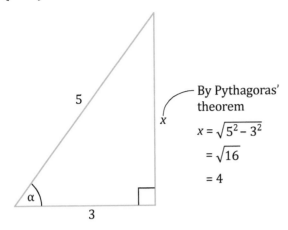

By Pythagoras' theorem

$$x = \sqrt{5^2 - 3^2}$$

$$= \sqrt{16}$$

$$= 4$$

Hence, $\sin \alpha = \frac{4}{5}$

Area $= \frac{1}{2} ab \sin C = \frac{1}{2} \times 21 \times 10 \times \frac{4}{5} = 84$ cm²

Examples

1 (a) Find all the values of θ in the range $0° \le \theta \le 360°$ satisfying $2 \sin \theta = 1$ [3]

  (b) Find all the values of θ in the range $0° \le \theta \le 2\pi$ satisfying $\tan \frac{\theta}{2} = \sqrt{3}$ giving your answers in terms of π. [3]

Notice that the first part of the question requires the angle in degrees and the other requires the angle in radians.

. . . .

Answer

1 (a) $2 \sin \theta = 1$

  so $\sin \theta = \frac{1}{2}$

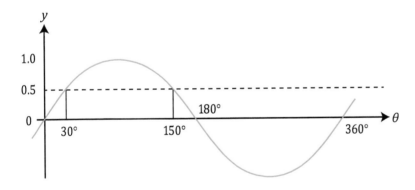

$\theta = \sin^{-1}\left(\frac{1}{2}\right)$

$\theta = 30°$

$\theta = 180 - 30 = 150°$

Hence $\theta = 30°$ or $150°$

(b) $\tan\dfrac{\theta}{2} = \sqrt{3}$

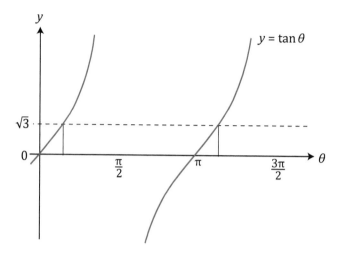

You could alternatively use the CAST method to find the solutions to $\tan\dfrac{\theta}{2} = \sqrt{3}$.

Using the CAST diagram, tan is positive in the first and third quadrants giving the values of:

$\dfrac{\theta}{2} = \dfrac{\pi}{3}$ or $\pi + \dfrac{\pi}{3} = \dfrac{4\pi}{3}$

You must remember to work in radians since the answers are required in terms of π.

$\dfrac{\theta}{2} = \tan^{-1}\sqrt{3}$

Since values of θ are asked for in the range $0° \le \theta \le 2\pi$, then $\dfrac{\theta}{2}$ will be in the range 0 to π, so there is no need to consider values greater than π.

$\tan\dfrac{\pi}{3} = \sqrt{3}$.
This result should be known.
Remember that $\dfrac{\pi}{3} = 60°$

Hence $\dfrac{\theta}{2} = \dfrac{\pi}{3}$

$\theta = \dfrac{2\pi}{3}$ radians

---

2   (a)   Find all values of θ between $0° \le \theta \le 360°$ satisfying
$8\cos^2\theta - 7\sin^2\theta = 4\cos\theta - 3$.   [6]

(b)   Find all values of x between $0°$ and $180°$ satisfying
$\cos(2x + 25°) = -0.454$.   [3]

. . . .

Answer

Although each of the terms in $\sin^2\theta$ or $\cos^2\theta$ could be written in terms of the other using $\sin^2\theta + \cos^2\theta = 1$, this is not as straightforward for the term in $\cos\theta$. Since there is a term in $\cos\theta$, we shall write the equation just in terms of $\cos\theta$.

2   (a) $8\cos^2\theta - 7\sin^2\theta = 4\cos\theta - 3$

$8\cos^2\theta - 7(1 - \cos^2\theta) = 4\cos\theta - 3$

$8\cos^2\theta - 7 + 7\cos^2\theta = 4\cos\theta - 3$

$15\cos^2\theta - 4\cos\theta - 4 = 0$

$(5\cos\theta + 2)(3\cos\theta - 2) = 0$

Using the CAST method cos is positive in the first and fourth quadrants so $\theta = 48.19°$ or $360° - 48.19° = 311.81°$. Alternatively you could draw a cos graph to help work out the values.

$\cos\theta = -\dfrac{2}{5}$ or $\dfrac{2}{3}$

When $\cos\theta = \dfrac{2}{3}$, $\theta = 48.19°$ or $311.81°$

When $\cos\theta = -\dfrac{2}{5}$, cos is negative in the second and third quadrants, so:
$\theta = 113.58°$
or $360 - 113.58 = 246.42°$

When $\cos\theta = -\dfrac{2}{5}$, $\theta = 113.58°$ or $246.42°$

$\theta = 48.19°, 113.58°, 246.42°, 311.81°$

(b) $\cos(2x + 25°) = -0.454$

$$2x + 25° = 117°, 243°$$

$$2x = 92°, 218°$$

$$x = 46°, 109°$$

Possible values of x are 46° and 109°

Since the question asks for all the solutions between 0° and 180°, and the angle involved in the question is $2x$, then values of $2x$ need to be considered between 0° and 360° only.

---

3 In triangle ABC, BC = 12 cm and $\cos A\widehat{B}C = \frac{2}{3}$.

The length of AC is 2 cm greater than the length of AB.

(a) Find the lengths of AB and AC. [4]

(b) Find the exact value of $\sin B\widehat{A}C$. Give your answer in its simplest form. [3]

Answer

3 (a) Let length of AB = x so AC = x + 2.

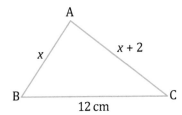

No diagram is given here so draw your own.

Using the cosine rule, we obtain

$$(x + 2)^2 = x^2 + 12^2 - 2 \times x \times 12 \times \frac{2}{3}$$

$$x^2 + 4x + 4 = x^2 + 144 - 16x$$

$$20x = 140$$

$$x = 7$$

Hence AB = 7 cm and AC = 9 cm

(b)

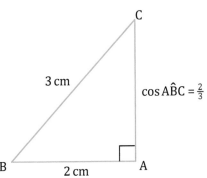

By Pythagoras' theorem $AC^2 = 3^2 - 2^2$, giving $AC = \sqrt{5}$

$$\sin A\widehat{B}C = \frac{\sqrt{5}}{3}$$

Applying the sine rule to the original triangle, we obtain

$$\frac{\sin B\widehat{A}C}{12} = \frac{\sin A\widehat{B}C}{9}$$

$$\sin B\widehat{A}C = \frac{4\sqrt{5}}{9}$$

BOOST

Grade ⬆⬆⬆⬆

Always check the question to see how it wants the answer. Here an exact value is needed so do not give the answer as a decimal.

Test yourself

1   The diagram shows the triangle ABC with AB = 8 cm, AC = 12 cm and
B$\hat{A}$C = 150°.

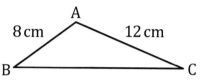

(a) Find the area of triangle ABC.

(b) Find the length of BC correct to one decimal place.

2   The graph shows the curve $y = \sin x$ in the interval $0 \leq x \leq 4\pi$

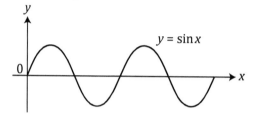

(a) Write down the coordinates of all the points of intersection with the x-axis.

(b) Write down the coordinates of all stationary points for this graph.

3   The triangle ABC is such that AB = 4 cm, BC = $(3\sqrt{2} - 1)$ cm and B$\hat{A}$C = 30°.

Find an expression for sin A$\hat{C}$B in the form $\frac{2 + m\sqrt{2}}{n}$, where m and n are integers whose values are to be found.

4   Find all the values of θ in the range $0° \leq \theta \leq 360°$ satisfying
$3 \sin^2 \theta = 5 - 5 \cos \theta$

5   The diagram below shows a sketch of the triangle ABC with AB = x cm,
AC = $(x - 2)$ cm and BC = $(x + 2)$ cm.

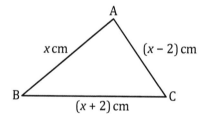

(a) Show that cos B$\hat{A}$C = $\frac{x - 8}{2x - 4}$.

(b) Given that B$\hat{A}$C = 120°, find the size of angle A$\hat{B}$C.

6   (a) Find all values of θ in the range $0° \leq x \leq 360°$ satisfying
$4 \cos^2 \theta + 1 = 4 \sin^2 \theta - 2 \cos \theta$

(b) The angle α satisfies

$$\sin(\alpha + 40°) = \frac{1}{\sqrt{2}}$$

and $\sin(\alpha - 35°) = \dfrac{\sqrt{3}}{2}$

Given that $0° \leq \alpha \leq 360°$, find the value of α.

(c) Find all values of ϕ in the range $0° \leq \phi \leq 360°$ satisfying

$$\frac{7}{\cos\phi} - \frac{10}{\sin\phi} = 0$$

Giving your answer to the nearest degree.

Summary

Trigonometry

The sine and cosine rules and the formula for the area of a triangle

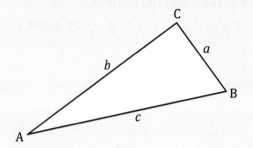

The sine rule states: $\dfrac{a}{\sin A} = \dfrac{b}{\sin B} = \dfrac{c}{\sin C}$

The cosine rule states: $a^2 = b^2 + c^2 - 2bc \cos A$

Area of a triangle $= \dfrac{1}{2} ab \sin C$

Trigonometric relationships

$\tan \theta = \dfrac{\sin \theta}{\cos \theta}$

$\cos^2 \theta + \sin^2 \theta = 1$

6 Exponentials and logarithms

Introduction

Exponentials and logarithms will probably be new to you but it's not as difficult as it sounds. Logarithms are often used to compare values which rise very quickly and are used for the decibel scale, which is a scale for the loudness of sound, and the Richter scale, which is used for the magnitude of earthquakes.

Make sure you have a thorough grasp of the laws of indices before starting this topic.

This topic covers the following:

6.1 $y = a^x$ and its graph

6.2 $y = e^x$ and its graph

6.3 The graph of $y = e^{kx}$ and the gradient = ke^{kx}

6.4 The definition of $\log_a x$ as the inverse of a^x

6.5 $\ln x$ and its graph

6.6 Proof and use of the laws of logarithms

6.7 Solving equations in the form $a^x = b$

6.8 Using exponential growth and decay in modelling

6.9 Limitations and refinements of exponential models

6.1 $y = a^x$ and its graph

The function $f(x) = a^x$, $a > 0$ is called an exponential function and different values of a will give different exponential functions. An important case uses $a = e$ where e is the special irrational value which is 2.718 to three decimal places. This is then called 'the exponential function'.

Don't get mixed up between a^x and e^x.

a^x is an **exponential function** and a can take many values.

e^x is a special case where $a = e$ and e^x is called **the exponential function**.

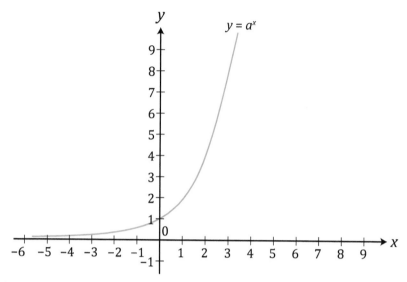

The graph of $y = a^x$ where a is a positive constant greater than 1, is shown above. Notice that the graph intersects the y-axis at 1. No matter what the positive value of a is, the intersect on the y-axis is always 1. The reason for this is that on the y-axis, $x = 0$ so $y = a^0 = 1$.

If the value of the positive constant a is less than 1, then the graph of $y = a^x$ looks like this.

When sketching these graphs ensure that the curve does not touch the x-axis.

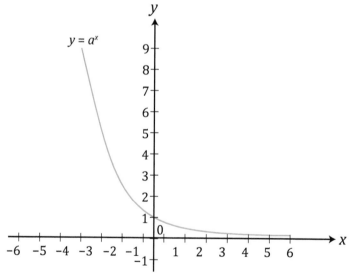

Notice that as the value of x decreases, the value of y approaches but never reaches zero. The line which a curve approaches but never touches is called an asymptote. The x–axis is an asymptote to the curve $y = a^x$.

It is important to note that the graphs of $y = 2^x$ and $y = 5^x$ (which both have the general form $y = a^x$) have the same general shapes but the larger the value of a (i.e. 2 and 5 in this case), the more steeply the graph rises as x increases.

6.2 *y* = eˣ and its graph

The graph of *y* = eˣ is a special case of the *y* = *aˣ* graph where e is the number 2.71828... .

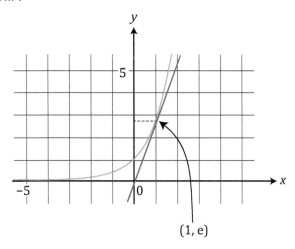

(1, e)

The main features of the graph *y* = eˣ are:

- Like all exponential graphs the curve is upward sloping and increases faster as *x* increases and also intersects the *y*-axis at *y* = 1.

- The graph lies above the *x*-axis but gets very near but never touches it for negative values of *x*. The *x*-axis is an asymptote.

- The gradient of the tangent to the graph at each point is equal to its *y*-coordinate at that point. Note that this applies only to the graph of *y* = eˣ.

The graph of *y* = eˣ is called **the exponential graph**.

Active Learning

Draw an accurate graph on graph paper of *y* = 3ˣ for −3 ≤ *x* ≤ 6. Use the table below to record the values of *x* and *y*.

Draw tangents to the graph at each of the points on the curve corresponding to the points *x* = −1, *x* = 0, *x* = 1, *x* = 2, *x* = 3, *x* = 4, *x* = 5.5 and then using these tangents, find the gradients (called *g* in the table below) of the curve at each of these points correct to one decimal place.

x	−1	0	1	2	3	4	5.5
y							
g							
$\frac{g}{y}$							

Complete the table above and comment on your results.

6.3 The graph of $y = e^{kx}$ and the gradient $= ke^{kx}$

This is a similar activity to the one in Section 6.2.

Draw an accurate graph on graph paper of $y = e^{2x}$ for $-3 \leq x \leq 6$. Use the table below to record the values of x and y.

Draw tangents to the graph at each of the points on the curve corresponding to the points $x = -1, x = 0, x = 1, x = 2, x = 3, x = 4, x = 5.5$ and then using these tangents, find the gradients (called g in the table below) of the curve at each of these points correct to one decimal place.

x	−1	0	1	2	3	4	5.5
y							
g							
$\dfrac{g}{y}$							

Complete the table above and comment on your results.

The gradient of the tangent to the curve $y = e^{kx}$

The gradient of the tangent to the curve $y = e^{kx}$ at each point is equal to the value of k multiplied by its y-coordinate at that point.

For example, the curve $y = e^{5x}$ would have a gradient of $5y$ or $5e^{5x}$.

6.4 The definition of $\log_a x$ as the inverse of a^x

What is a logarithm?

A logarithm of a positive number to a base a is the power to which the base must be raised in order to give the positive number. We can write two important equations using this definition where the positive number is y, the base is a and the log of the number is x:

$$y = a^x$$
$$\log_a y = x$$

Both of these equations have the same meaning and you must be able to convert between them.

The following example will help explain this:

If $y = 10^3$ then from the second equation we have $\log_{10} y = 3$

You must remember both of these equations and be able to use them.

For a positive base a, the following are true:

$$\log_a a = 1, \text{ as } a^1 = a.$$
$$\log_a 1 = 0, \text{ as } a^0 = 1.$$

The definition of $\log_a x$ as the inverse of a^x

The rules of logarithms are covered in detail in Section 6.6.

The inverse of an exponential function is called its logarithm. For example, if $y = a^x$ and this is substituted into $\log_a y$ you obtain:

$\log_a a^x = x \log_a a$ and as $\log_a a = 1$, $x \log_a a = x$.

This means that you get back to x the original value. So the inverse of a^x is $\log_a x$ and vice versa.

To solve the equation $\log (2x + 1) = \log x$, we can take exponentials of both sides which removes the log to give $2x + 1 = x$, giving $x = -1$.

If the special value e is substituted for a, then we have the inverse of e^x as $\log_e x$ which is usually written as $\ln x$.

Examples

1   Solve the equation $\ln (3x - 5) = \ln 2x$.

. . . .

Answer

1    $\ln (3x - 5) = \ln 2x$

Taking exponentials of both sides

$\quad 3x - 5 = 2x$

$\quad\quad x = 5$

Taking exponentials of both sides removes the logarithms.

---

2   Solve $\log (x^2 - 3x - 2) = \log (x + 3)$

. . . .

Answer

2   Taking exponentials of both sides

$\quad x^2 - 3x - 2 = x + 3$

$\quad x^2 - 4x - 5 = 0$

$\quad (x - 5)(x + 1) = 0$

$\quad\quad\quad x = 5$ or -1

6.5 ln x and its graph

The inverse of a function will produce the input value from the output value. The inverse of $y = e^x$ is $y = \ln x$. The graph of $\ln x$ and its inverse function e^x are shown below.

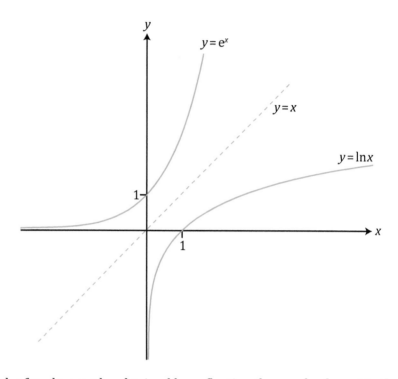

When reflecting in the line, you need to ensure that the scales on both axes are the same.

The graph of $y = \ln x$ can be obtained by reflecting the graph of $y = e^x$ in the line $y = x$

Notice also that:

The x-axis is an asymptote to the curve $y = e^x$.

The y-axis is an asymptote to the curve $y = \ln x$.

Since the action of an inverse function reverses the effect of the function, then two useful results follow: $e^{\ln x} = x$ and $\ln e^x = x$.

6.6 Proof and use of the laws of logarithms

The following proofs of the laws of logarithms need to be remembered for the examination. You will also be required to use these laws.

Proving $\log_a x + \log_a y = \log_a (xy)$

Suppose $x = a^n$ and $y = a^m$

These can be rewritten as:

$$\log_a x = n \text{ and } \log_a y = m$$
$$xy = a^n \times a^m$$
$$xy = a^{n+m}$$

Remember that according to the laws of indices the powers here are added.

This can be rewritten as:

$$\log_a(xy) = n + m$$

But $n = \log_a x$ and $m = \log_a y$

$$\boxed{\log_a x + \log_a y = \log_a(xy)}$$

Proving $\log_a x - \log_a y = \log_a(x/y)$

Suppose $x = a^n$ and $y = a^m$

These can be rewritten as:

$$\log_a x = n \text{ and } \log_a y = m$$

Now

$$\frac{x}{y} = \frac{a^n}{a^m}$$

$$\frac{x}{y} = a^{n-m}$$

> Remember that according to the laws of indices the powers here are subtracted.

These can be rewritten as:

$$\log_a \frac{x}{y} = n - m$$

But $n = \log_a x$ and $m = \log_a y$

Hence

$$\boxed{\log_a x - \log_a y = \log_a \frac{x}{y}}$$

Proving $k \log_a x = \log_a(x^k)$

Suppose $x = a^n$, then $\log_a x = n$

Raising both sides of the equation to the power k gives

$$x^k = (a^n)^k$$

$$x^k = a^{nk}$$

> According to the laws of indices, you multiply the powers inside and outside the brackets.

This can be rewritten as:

$$\log_a x^k = nk$$

As $n = \log_a x$ this can be substituted in for n.

Hence

$$\boxed{\log_a x^k = k \log_a x}$$

All three laws can be used when dealing with expressions or equations involving logs. Examples involving the simplification of expressions containing logs are shown here.

Examples

1 Express $\log_a 64 - 2\log_a 4$ as a single logarithm in the form $\log_a b$ where b is an integer.

· ·

Answer

1 $\log_a 64 - 2\log_a 4 = \log_a 64 - \log_a 4^2$

$$= \log_a 64 - \log_a 16$$

$$= \log_a \left(\frac{64}{16}\right)$$

$$= \log_a 4$$

---

2 Express $\frac{1}{2}\log_a 9 + \log_a 3 - 3\log_a 3$ as a single term.

· ·

Answer

2 $\frac{1}{2}\log_a 9 + \log_a 3 - 3\log_a 3 = \log_a 9^{\frac{1}{2}} + \log_a 3 - 3\log_a 3$

$$= \log_a 3 + \log_a 3 - \log_a 27$$

$$= \log_a \left(\frac{3 \times 3}{27}\right)$$

$$= \log_a \left(\frac{1}{3}\right)$$

$$= \log_a 1 - \log_a 3$$

$$= -\log_a 3$$

> Here we are using this law of logarithms $\log_a x^k = k\log_a x$

> Here we are using this law of logarithms
> $\log_a x - \log_a y = \log_a \left(\frac{x}{y}\right)$

> Note that $9^{\frac{1}{2}} = \sqrt{9} = \pm 3$ As you cannot have the logarithim of a negative number, only the positive value is used.

6.7 Solving equations in the form $a^x = b$

Equations in the form $a^x = b$ can be solved by first taking logs to base a of both sides like this:

$$a^x = b$$

$$x = \log_a b$$

> $\log_a a^x = x\log_a a = x$
> as $\log_a a = 1$

Alternatively, you should recognise these two equations as having the same meaning.

Step by STEP

Solve $\log_a 90x^2 - \log_a \left(\frac{5}{x}\right) = \frac{1}{2}\log_a 144x^8$

Steps to take

1 Combine the two log terms on the left to give a single log term, and cancel if possible.

2 On the right-hand side, raise the $144x^8$ to the power $\frac{1}{2}$ and simplify by square rooting $144x^8$.

3 Remove the log and solve the resulting equation for x.

Answer

$$\log_a 90x^2 - \log_a \left(\frac{5}{x}\right) = \tfrac{1}{2} \log_a 144x^8$$

$$\log_a 90x^2 \left(\frac{x}{5}\right) = \log_a (144x^8)^{\frac{1}{2}}$$

$$\log_a 18x^3 = \log_a 12x^4$$

$$18x^3 = 12x^4$$

$$x = 1.5$$

Remember the rules of indices to simplify $(144x^8)^{\frac{1}{2}}$

To remove the logs, take exponentials of both sides.

Examples

1 Solve the equation $\log_4 x = -\tfrac{1}{2}$

Answer

1 $\log_4 x = -\tfrac{1}{2}$

$$x = 4^{-\frac{1}{2}} = \frac{1}{4^{\frac{1}{2}}} = \frac{1}{\sqrt{4}}$$

$$x = \pm\tfrac{1}{2}$$

When taking a square root, you must remember to include the ±; however, you cannot find the logarithm of a negative number so the negative solution is ignored here.

x cannot be negative so $x = \tfrac{1}{2}$

BOOST

Grade ⬆⬆⬆⬆

Ensure you are fully confident in converting from log to exponential form and vice versa.

2 Solve the equation $2^{2x-1} = 9$
giving your answer correct to three decimal places.

Questions like this are solved by taking logarithms of both sides. The base used here is base 10. If the base is not shown, it is assumed that it is base 10.

Answer

2 $2^{2x-1} = 9$

$$\log 2^{2x-1} = \log 9$$

$$(2x - 1)\log 2 = \log 9$$

$$2x - 1 = \frac{\log 9}{\log 2} = 3.1699$$

$$2x = 4.1699$$

$$x = 2.085 \quad (3 \text{ d.p.})$$

Use the Log button on your calculator to work out the logs of numbers. Note that here:
$\frac{\log 9}{\log 2}$ does **not** equal $\log \frac{9}{2}$

When giving a numerical answer always check whether the answer needs to be given to a certain number of significant figures or decimal places.

3 Solve the equation $\log_a (3x + 4) = \log_a 5 + \log_a x$

Answer

3 $\log_a (3x + 4) = \log_a 5 + \log_a x$

$$\log_a (3x + 4) = \log_a 5x$$

$$3x + 4 = 5x$$

$$4 = 2x$$

$$x = 2$$

4   Solve the equation $25^x - 4 \times 5^x + 3 = 0$ where $x > 0$

You need to recognise that this equation is similar in format to a quadratic equation.

Answer

4    $25^x - 4 \times 5^x + 3 = 0$

Now $25^x = (5^2)^x = 5^{2x} = (5^x)^2$

Notice that

$$25^x = (5^2)^x = 5^{2x} = (5^x)^2$$

Hence $5^{2x} - 4 \times 5^x + 3 = 0$

So $(5^x)^2 - 4 \times 5^x + 3 = 0$

Let $y = 5^x$

Notice that $25^x = 5^{2x}$. The substitution $y = 5^x$ is used to obtain a quadratic equation in y which can then be factorised and solved.

$$y^2 - 4y + 3 = 0$$

$$(y - 1)(y - 3) = 0$$

$y = 1$ or $y = 3$

When $y = 1$, $1 = 5^x$

$$5^0 = 1 \quad \text{so } x = 0$$

We cannot have this value as x must be greater than 0.

When $y = 3$

$$5^x = 3$$

Taking logs to base 10 of both sides

$$\log 5^x = \log 3$$

$$x \log 5 = \log 3$$

$$x = \frac{\log 3}{\log 5}$$

$$x = 0.68 \quad \text{(2 d.p.)}$$

When there is more than one value for x, always check whether each value is possible. Look back at the question to see if there is a restriction on x. Here the restriction is $x > 0$

5   Given that $\log_d z = 2 \log_d 6 - \log_d 9 - 1$

Find z in terms of d

Answer

5    $\log_d z = 2 \log_d 6 - \log_d 9 - 1$

$\log_d z = \log_d 36 - \log_d 9 - \log_d d$

$\log_d z = \log_d \dfrac{36}{9d}$

$\log_d z = \log_d \dfrac{4}{d}$

$z = \dfrac{4}{d}$

$\log_d d = 1$

6   Solve the equation $\log_a (6x^2 + 9x + 2) - \log_a x = 4 \log_a 2$

Answer

6    $\log_a (6x^2 + 9x + 2) - \log_a x = 4 \log_a 2$

$$\log_a \left(\frac{6x^2 + 9x + 2}{x} \right) = \log_a 2^4$$

$$\log_a \left(\frac{6x^2 + 9x + 2}{x} \right) = \log_a 16$$

$$\frac{6x^2 + 9x + 2}{x} = 16$$

$$6x^2 - 7x + 2 = 0$$

$$(3x - 2)(2x - 1) = 0$$

$$x = \frac{2}{3} \text{ or } \frac{1}{2}$$

---

7   (a)   Given that $x > 0$ show that $\log_a x^n = n \log_a x$

(b)   Express $\frac{1}{2} \log_a 324 + \log_a 56 - 2 \log_a 12$ in the form $\log_a b$, where b is a constant whose value is to be found.   [4]

(c)   (i)   Rewrite the equation $3^x = 2^{x+1}$ in the form $c^x = d$ where the values of the constants c and d are to be found.

(ii)   Hence or otherwise, solve the equation $3^x = 2^{x+1}$ giving your answer correct to two decimal places.   [4]

Answer

7   (a)   See proof on page 121.

(b) $\frac{1}{2} \log_a 324 + \log_a 56 - 2 \log_a 12 = \log_a 324^{\frac{1}{2}} + \log_a 56 - \log_a 12^2$

$$= \log_a 18 + \log_a 56 - \log_a 144$$

$$= \log_a \left(\frac{18 \times 56}{144} \right) = \log_a 7$$

$324^{\frac{1}{2}} = \sqrt{324} = 18$

(c)   (i) $3^x = 2^{x+1}$

$3^x = 2^x \times 2^1$ Notice the way the indices are separated here. The laws of indices are used here to separate 2^{x+1} into $2^x \times 2^1$

$$\left(\frac{3^x}{2^x} \right) = 2$$

$$\left(\frac{3}{2} \right)^x = 2$$

Hence $c = \frac{3}{2}$ and $d = 2$.

(ii) $\left(\frac{3}{2} \right)^x = 2$

Look back at the question to check if you have the answer in the correct format. Here we need an answer in the form $c^x = d$.

The answer is requested in this format with $c = \frac{3}{2}$ and $d = 2$

Taking logs of both sides:

$$\log \left(\frac{3}{2} \right)^x = \log 2$$

$$x = \frac{\log 2}{\log \left(\frac{3}{2} \right)} = \frac{0.3010}{0.1761} = 1.71 \quad (2 \text{ d.p.})$$

When you have an answer, always check whether it needs to be given to a certain number of decimal places or significant figures.

6.8 Using exponential growth and decay in modelling

The function e^x can be used to model real situations which have exponential growth such as population growth, growth of bacteria etc. and the function e^{-x} can model radioactive decay.

Two common types of mathematical models are

Exponential growth: $y = ae^{kt}$, $k > 0$.

Exponential decay: $y = ae^{-kt}$, $k > 0$.

> Exponential decay or growth crops up in many situations in chemistry, physics and biology.

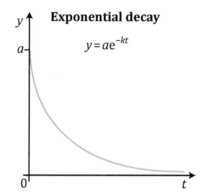

Exponential decay

$y = ae^{-kt}$

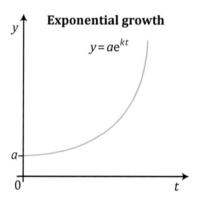

Exponential growth

$y = ae^{kt}$

Notice both of these graphs cut the y-axis at $y = a$ which is the y-value at $t = 0$ (i.e. the initial value of the quantity being modelled).

Example

1 The growth of an investment of £P at a rate of compound interest of r% over a time of t years can be modelled using the formula

$$A = Pe^{rt}$$

If £1000 is invested at a continuous rate of compound interest of 5%, how long will it take for the investment to double?

Answer

1   Using $A = Pe^{rt}$, we have $A = 1000e^{0.05t}$

We want to find t when A doubles so we set A equal to 2000 and solve for t.

$1000e^{0.05t} = 2000$. Divide both sides by 1000.

$e^{0.05t} = 2$

Taking ln of both sides we obtain

$\ln e^{0.05t} = \ln 2$

$0.05t = \ln 2$

$t = \dfrac{\ln 2}{0.05}$

$= 13.86$

$= 14$ years (as t needs to be an integer)

> Notice that as the index is positive this represents exponential growth.

> Note that $\ln e = 1$.

Step by STEP

The value £V of a long-term investment may be modelled using the formula

$V = Ae^{kt}$ where A and k are constants

The value of the investment after 2 years is £292 and its value after 28 years is £637.

Find the initial value of the investment; give your answer correct to the nearest pound.

Steps to take

1   Note that there are two constants in this formula, A and k, and before we can use this formula to find V, we need to find their value.

2   Notice that two sets of pairs of values are given. When this happens you usually substitute them into the formula in turn to produce a pair of equations which can be solved simultaneously.

3   Once the constants have been found they are substituted back into the formula and then the formula can be used to find other values.

BOOST
Grade ⇧⇧⇧⇧

> Look out for simultaneous equations when two sets of conditions are given.

Answer

When $t = 28$, $V = 637$ so $637 = Ae^{28k}$

When $t = 2$, $V = 292$ so $292 = Ae^{2k}$

Dividing these two equations, we obtain $\dfrac{637}{292} = \dfrac{Ae^{28k}}{Ae^{2k}}$

$2.1815 = e^{26k}$

Taking ln of both sides $\ln 2.1815 = \ln e^{26k}$

$\ln 2.1815 = 26k$

$k = \dfrac{\ln 2.1815}{26} = 0.03$

Now $637 = Ae^{28k}$, so $637 = Ae^{28 \times 0.03}$ giving $A = 275$

When $t = 0$, $V = 275e^{0} = 275$

Initial investment = £275

> Dividing the equations eliminates A.

6.9 Limitations and refinements of exponential models

Simple exponential models provide a way of understanding how a real situation occurs. Usually they only mimic the real situation for a certain range of values. For example, the new customers joining a new social media website may follow an exponential curve with time over a limited time period but this will not carry on forever. Likewise, population growth in a new city may be approximately exponential at first but reach a steady population after a certain time.

You must be able to discuss the limitations of any exponential model.

Examples

1 A population, P, of a town is growing exponentially and can be modelled using the formula $P = Ae^{kt}$, where t is measured in years.

> If the population is 15 000 initially, this is the population of the town when $t = 0$.

    (a) If $P = 15\,000$ at the start of year 2000, and P has grown to 17 000 at the start of year 2003, find the formula for P.

    (b) Based on this model, at the end of which year will the population reach 20 000?

· ·

Answer

1   (a) When $t = 0$, $P = 15\,000$ so using $P = Ae^{kt}$ we have $15\,000 = Ae^0$ and as $e^0 = 1$ the constant $A = 15\,000$.

          We need to find k. We know that k is positive, since the population is growing. Using the value we have found for A, we have

$$P = 15000\,e^{kt}.$$

          The year 2003 corresponds to $t = 3$, so we substitute $P = 17\,000$ and $t = 3$ in the equation above and solve for k.

$$17\,000 = 15\,000\,e^{3k}$$

$$e^{3k} = \frac{17\,000}{15\,000}$$

$$e^{3k} = \frac{17}{15}$$

          Taking ln of both sides, we obtain

$$\ln\left(\frac{17}{15}\right) = \ln e^{3k}$$

> $\ln e^{3k} = 3k$ as ln and e are inverse functions.

$$3k = \ln\left(\frac{17}{15}\right)$$

$$k = 0.0417 \text{ (3 s.f.)}$$

          Hence $P = 15\,000e^{0.0417t}$

    (b) Based on this model, when will the population reach 20 000?

          When $P = 20\,000$, $20\,000 = 15\,000e^{0.0417t}$

$$e^{0.0417t} = \frac{20\,000}{15\,000}$$

$$e^{0.0417t} = \frac{4}{3}$$

Taking ln of both sides,

$$\ln(e^{0.0417t}) = \ln\left(\frac{4}{3}\right)$$

$$0.0417t = \ln\left(\frac{4}{3}\right)$$

$$t = 6.9 \text{ years}$$

So if the model is correct, the population would have reached 20 000 at the end of 2006.

> When you find the ln of an exponential you just get the power to which the exponential is raised.

2   The number of undecayed radioactive atoms after t years can be modelled using the formula

$$N = N_0 e^{-kt}$$

(a)   State the significance of the value N_0.

(b)   It is known that the number of undecayed radioactive atoms remaining after 5600 years is half that of the number originally present.

Find the value of k correct to 3 significant figures.

(c)   A sample contains 1×10^{20} undecayed atoms at $t = 0$.

How many undecayed atoms will there be after 1000 years? Give your answer in standard form correct to 3 significant figures.

. . . .

Answer

2   (a) $N = N_0 e^{-kt}$ and when $t = 0$, $N = N_0$ as $e^0 = 1$

N_0 is the number of undecayed atoms initially present (i.e. at $t = 0$)

(b)
$$N = \frac{N_0}{2}$$

$$\frac{N_0}{2} = N_0 e^{-kt}$$

$$\frac{1}{2} = e^{-k \times 5600}$$

$$\ln\left(\frac{1}{2}\right) = -5600k$$

$$k = 0.000\,124$$

> The number of undecayed atoms after 5600 years is half the number originally present $\left(\text{i.e. } \frac{N_0}{2}\right)$

(c)
$$N = 1 \times 10^{20} e^{-0.000\,124 \times 1000}$$

$$N = 8.83 \times 10^{19} \text{ (3 s.f.)}$$

BOOST

Grade ⇧⇧⇧⇧

Whenever you write an answer, look back at the question to see if it specifies a certain number of decimal places or significant figures.

3   The number of regular users of a certain website has grown exponentially over the last 5 years and can be modelled using the formula

$$A = Ne^{0.6t}$$

There were 300 000 users when the website went live.

(a)   Find how many users there are after 5 years. Give your answer to 1 s.f.

(b)   Find how many users there are after 20 years. Give your answer to 1 s.f.

(c)   Using your answer from part (b), comment on the validity of the model.

Answer

3   (a) $A = Ne^{0.6t}$

$A = 300\,000e^{0.6 \times 5} = 6\,025\,661 = 6\,000\,000$ (1 s.f.)

(b) $A = Ne^{0.6t}$

$A = 300\,000e^{0.6 \times 20} = 5 \times 10^{10}$ (1 s.f.)

(c)   The answer is too large so the exponential model is only valid for a smaller number of years. The eventual users are limited by the size of the population.

> Remember that models often do not work properly in every situation. Always ask yourself if an answer seems reasonable/sensible.

6.10 Using logarithmic graphs to reduce exponential equations to linear form

Relationships such as $y = ax^n$ and $y = kb^x$ are called exponential relationships and they often arise when you are trying to model a particular real-life situation such as population growth or the variation of quantities in a science experiment.

The values of a, b and k are called parameters and we often have to find their values when there are known values for x and y. To find the values of these parameters it is much easier to turn the equations of curves into equations of straight lines. It is much easier to work with the equation of a straight line as it is easy to find the intercept and gradient and this allows us to determine the values of unknown parameters.

If the relationship is $y = ax^n$ we can turn this equation into that of a straight line in the following way:

$$y = ax^n$$

> Remember that
> $\log ab = \log a + \log b$
> and $\log x^n = n \log x$

Taking logarithms of both sides,	$\log y = \log(ax^n)$
The log on the right can be separated, so	$\log y = \log a + \log x^n$
The power n, can be placed in front of the log to give	$\log y = \log a + n \log x$
We can rearrange this to give	$\log y = n \log x + \log a$

Now this can be compared with the equation for a straight line $y = mx + c$

So if $\log y$ is plotted on the y-axis and $\log x$ is plotted on the x-axis then the graph of $\log y = n \log x + \log a$ will be a straight line with gradient n and intercept on the y-axis $\log a$.

Hence it is now possible to find the values of the parameters a and n.

6.10 Using logarithmic graphs to reduce exponential equations to linear form

In a similar way, the equation $y = kb^x$ can be turned into the equation of a straight line in the following way:

Note that when drawing this graph, $\log y$ on the y-axis is plotted against x on the x-axis.

Taking logarithms of both sides we obtain $\qquad\qquad \log y = \log k + \log b^x$

Now putting the index x in front of the log we obtain $\quad \log y = \log k + x \log b$

Rearranging this equation $\qquad\qquad\qquad \log y = x \log b + \log k$

This can now be compared with the equation of a straight line, $y = mx + c$

You can see the $\log y$ can be compared with the y, the x is the same in both equations, the $\log b$ can be compared with the m and $\log k$ is c.

Hence if we plot $\log y$ on the y-axis and x on the x-axis then the gradient will be equal to $\log b$ and the intercept on the y-axis will equal $\log k$.

Suppose we are told that the x and y values in the following table are connected by the equation $y = ax^n$ where a and n are integers.

x	2	30	70	100	150
y	4.24	16.4	25.1	30.0	36.7

To find the values of a and n we take logarithms of both sides and simplify and rearrange the equation until it is in the form $y = mx + c$

On doing this we obtain $\qquad \log y = n \log x + \log a$

You can compare this equation with the standard equation of a straight line so y is $\log y$, m is n, x is $\log x$ and c is $\log a$.

This equation will produce a straight line when $\log x$ is plotted on the x-axis and $\log y$ is plotted on the y-axis. We now need to produce a new table showing the values of $\log x$ and $\log y$.

$\log x$	0.30	1.48	1.85	2.00	2.18
$\log y$	0.63	1.21	1.40	1.48	1.56

We now plot the graph with $\log y$ on the y-axis and $\log x$ on the x-axis.

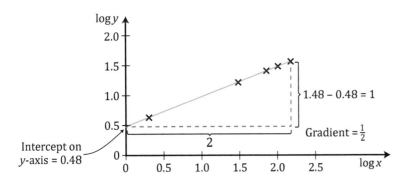

The intercept on the y-axis is read off from the graph and found to be 0.48.

From the equation of the line $\log y = n \log x + \log a$ we obtain $\log a = 0.48$

Now $a = 10^{0.48}$ so $a = 3.0$ (2 s.f.)

The gradient of the graph is found by drawing a suitable triangle.

Now gradient $= \frac{1}{2}$ so $n = \frac{1}{2}$

Substitute the values for a and n into the original equation for the curve of $y = ax^n$.

This means that the original equation of the curve was $y = 3x^{\frac{1}{2}}$

Example

1   When an object moves through air it experiences air resistance. The air resistance R newtons is thought to obey the equation $R = av^b$

Where v is the speed in m s^{-1} and R is the resistance in newtons and a and b are constants.

An experiment was conducted and the following results were obtained.

v	10	20	30	40
R	73	260	545	920

(a)   Show that the equation $R = av^b$ can be written as $\log R = b \log v + \log a$.

(b)   By plotting a suitable graph, find the values of a and b, giving your answers to two significant figures.

(c)   Find the value of R when $v = 5$ m s^{-1}. Give your answer to two significant figures.

. . . .

Answer

1   (a) $R = av^b$

$\log R = \log av^b$

$\log R = \log a + \log v^b$

> This equation is now in the form $y = mx + c$.

$\log R = \log a + b \log v$

$\log R = b \log v + \log a$

(b)

$\log v$	1.00	1.30	1.48	1.60
$\log R$	1.86	2.41	2.74	2.96

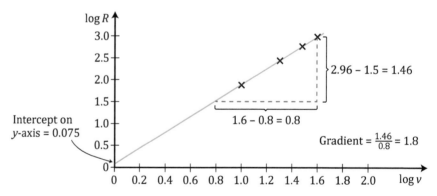

$\log R = b \log v + \log a$

From the graph, gradient $b = 1.825 = 1.8$ (2 s.f.)

From the graph, intercept on the y-axis, $\log a = 0.075$

Now $a = 10^{0.075} = 1.1885 = 1.2$ (2 s.f.)

Hence $a = 1.2$ and $b = 1.8$ (both to 2 s.f.)

(c)   Substituting the values of a and b into the formula we obtain $R = 1.2v^{1.8}$

When $v = 5$, $R = 1.2 \times 5^{1.8} = 21.7433 = 22$ N (2 s.f.)

Test yourself

1 Simplify $\log_2 36 - 2 \log_2 15 + \log_2 100$
expressing your answer in the form $\log_2 a$ where a is an integer.

2 Solve the equation $\log_{27} x = \frac{2}{3}$

3 Solve the equation $3^x = 2$
giving your answer correct to two decimal places.

4 Express $\frac{1}{2} \log_a 36 - 2 \log_a 6 + \log_a 4$ as a single logarithm.

5 Solve the equation $\log_a (6x^2 + 5) - \log_a x = \log_a 17$

6 Prove that: $\log_7 a \times \log_a 19 = \log_7 19$
whatever value of the positive constant a.

7 Solve $6^x = 12$, giving your answer correct to three decimal places.

8 Solve the equation $9^x - 6 \times 3^x + 8 = 0$ where $x > 0$
giving x correct to two decimal places.

9 Find all the values of x satisfying the equation
$\log_a (6x^2 + 11) - \log_a x = 2 \log_a 5$ [5]

10 (a) Given that $x > 0$, show that $\log_a x^n = n \log_a x$ [3]
(b) Solve the equation $6^{2y-1} = 4$
Show your working and give your answer correct to three decimal
places. [3]
(c) Given that $\log_a 4 = \frac{1}{2}$, find the value of a. [2]

11 The size N of the population of a small island at time t years may be
modelled by Ae^{kt}, where A and k are constants. It is known that $N = 100$
when $t = 2$ and that $N = 160$ when $t = 12$.
(a) Interpret the constant A in the context of the question. [1]
(b) Show that $k = 0.047$, correct to three decimal places. [4]
(c) Find the size of the population when $t = 20$. [3]

12 Given that $x > 0, y > 0$:
(a) Show that $\log_a xy = \log_a x + \log_a y$

(b) Solve the equation $2^{3-5x} = 12$
Show your working and give your answer correct to three decimal places.

(c) Solve the equation
$\log_9 (3x - 1) + \log_9 (x + 4) - 2 \log_9 (x + 1) = \frac{1}{2}$

Summary

The graphs of exponential and logarithmic functions

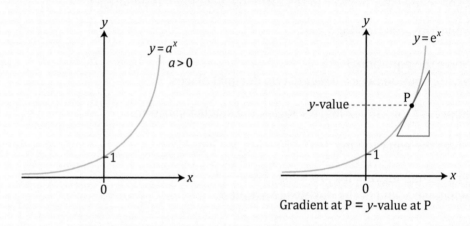

Gradient at P = y-value at P

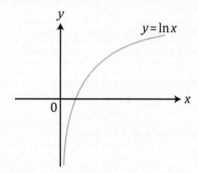

Logarithms and their uses

The logarithm and exponential functions

A logarithm of a positive number to base a is the power to which the base must be raised in order to give the positive number.

$$y = a^x$$

$$\log_a y = x$$

These two equations have the same meaning and you should be able to convert readily between them.

Some important results

For a positive base a, the following are true:

$$\log_a a = 1, \quad \text{as } a^1 = a$$

$$\log_a 1 = 0, \quad \text{as } a^0 = 1$$

The three laws of logarithms

$$\log_a x + \log_a y = \log_a (xy)$$

$$\log_a x - \log_a y = \log_a \frac{x}{y}$$

$$\log_a x^k = k \log_a x$$

7 Differentiation

Introduction

Differentiation can be used to find the gradient at a certain point on a curve or it can be used to determine the coordinates of stationary points where the gradient of the curve is zero.

Differentiation is also an important tool for finding maximum and minimum values of functions.

7.1 What is differentiation?

Unlike a straight line, which has a fixed gradient, a curve has a variable gradient depending on the point on the curve where the gradient is taken. The gradient at a point of the curve is the gradient of the tangent to the curve at that point. A tangent is a straight line that touches the curve at a point P (x, y).

The gradient of a curve depends on where it is taken. You will usually be given the x-coordinate where the gradient is to be found.

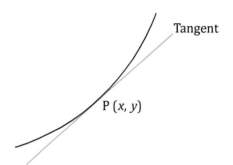

Differentiation is the process of finding a general expression for the gradient of a curve at any point. This general expression for the gradient is known as the derivative, and can be expressed in two ways: $\dfrac{dy}{dx}$ or $f'(x)$.

7.2 Differentiation from first principles

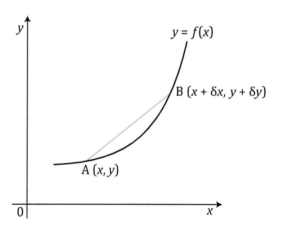

The line joining points A and B is called a chord. Notice that there is a small horizontal distance δx and a small vertical distance δy between points A and B. As A and B move closer together the gradient of the chord AB becomes nearer the true gradient of the tangent to the curve at point A. As $\delta x \to 0$ (i.e. as δx approaches zero) the chord will tend to become the tangent to the curve at point A and the gradient of the curve at point A will be the limit of the gradient of the chord.

This can be expressed in the following way:

$$\frac{dy}{dx} = \lim_{\delta x \to 0} \frac{\delta y}{\delta x} = \lim_{\delta x \to 0} \left(\frac{f(x + \delta x) - f(x)}{\delta x} \right)$$

You have to be able to differentiate from first principles a polynomial of degree less than three (i.e. up to and including terms in x^2).

Suppose we want to find $\dfrac{dy}{dx}$ for $y = 4x^2 - 2x + 1$

Increasing x by a small amount δx will result in y increasing by a small amount δy.

Substituting $x + \delta x$ and $y + \delta y$ into the equation we have:

$$y + \delta y = 4(x + \delta x)^2 - 2(x + \delta x) + 1$$

$$y + \delta y = 4(x^2 + 2x\delta x + (\delta x)^2) - 2x - 2\delta x + 1$$

$$y + \delta y = 4x^2 + 8x\delta x + 4(\delta x)^2 - 2x - 2\delta x + 1$$

But $\quad y = 4x^2 - 2x + 1$

Subtracting these equations gives

$$\delta y = 8x\delta x + 4(\delta x)^2 - 2\delta x$$

Dividing both sides by δx

$$\frac{\delta y}{\delta x} = 8x + 4\delta x - 2$$

Letting $\delta x \rightarrow 0$

$$\frac{dy}{dx} = \underset{\delta x \rightarrow 0}{\text{limit}} \frac{\delta y}{\delta x} = 8x - 2$$

7.3 Differentiation of x^n and related sums and differences

Before differentiating an expression, it needs to be written in index form. You may need to look back at Topic 1 to revise indices.

To differentiate an expression: multiply by the index and then reduce the index by one.

If $y = kx^n$ then the derivative $\dfrac{dy}{dx} = nkx^{n-1}$

Examples

1 If $y = 6x^3 + \frac{1}{2}x^2 - 5x + 4$, find $\dfrac{dy}{dx}$.

....

Answer

1 Differentiating gives

$$\frac{dy}{dx} = (3)6x^2 + (2)\tfrac{1}{2}x - 5$$

When differentiating a term in x you obtain the coefficient of x (e.g. $5x$ differentiated becomes 5). A number on its own when differentiated becomes zero.

$$\frac{dy}{dx} = 18x^2 + x - 5$$

2 Find the gradient of the curve $y = 3x^2 - x + 2$ at the point P (2, 12).

....

Answer

2 Differentiating the equation of the curve gives

$$\frac{dy}{dx} = (2)3x^1 - 1 = 6x - 1$$

At P, $x = 2$ so gradient $\dfrac{dy}{dx} = (6)2 - 1 = 11$

BOOST

Grade ⇧⇧⇧⇧

You must include this step about the limits. Leaving this step out will cost you a mark.

Active Learning

Differentiation from first principles is fairly easy but it takes a bit of practice.

Every few weeks make up a simple expression such as $9x^2 - 8x - 3$ and differentiate it from first principles.

BOOST

Grade ⇧⇧⇧⇧

This step shows the working.

You should show your working because if you make an arithmetic error, then you may still get marks for your method.

The x value of the point on the curve where the gradient is to be found is substituted into the expression for $\dfrac{dy}{dx}$

3 Given that $y = \sqrt{x} + \dfrac{4}{x^3} + 4$, find the value of $\dfrac{dy}{dx}$ when $x = 1$.

Answer

3 Writing the equation in index form gives

$$y = x^{\frac{1}{2}} + 4x^{-3} + 4$$

Differentiating gives

$$\frac{dy}{dx} = \frac{1}{2}x^{-\frac{1}{2}} + (-3)4x^{-4} = \frac{1}{2}x^{-\frac{1}{2}} - 12x^{-4}$$

> Be careful here as a common mistake is to convert to index form and then forget to differentiate the result.

Writing this in a form in which numbers are easily entered gives

$$\frac{dy}{dx} = \frac{1}{2\sqrt{x}} - \frac{12}{x^4}$$

Substituting $x = 1$ gives

$$\frac{dy}{dx} = \frac{1}{2\sqrt{1}} - \frac{12}{1^4} = \frac{1}{2} - 12 = -11.5$$

> When differentiating a term with a negative index, the index of the derivative will still be one less, e.g. if $y = x^{-3}$ then $\dfrac{dy}{dx} = -3x^{-4}$

> The derivative of $f(x)$ is written as $f'(x)$.

4 If $f(x) = \dfrac{3}{4}x^{\frac{1}{3}} - \dfrac{12}{x^2}$, find the value of $f'(x)$ when $x = 8$.

Answer

4 Writing the whole function in index form gives

$$f(x) = \frac{3}{4}x^{\frac{1}{3}} + 12x^{-2}$$

> You need to be confident in reducing fractional indices by 1 $\left(\text{e.g. } \frac{1}{2} - 1 = -\frac{1}{2}, \quad -\frac{1}{2} - 1 = -\frac{3}{2}, \text{ etc.}\right)$

Differentiating gives

$$f'(x) = \left(\frac{1}{3}\right)\frac{3}{4}x^{-\frac{2}{3}} + (-2)12x^{-3} = \frac{1}{4}x^{-\frac{2}{3}} + -24x^{-3}$$

Writing this in non-index form gives:

$$f'(x) = \frac{1}{4\sqrt[3]{x^2}} - \frac{24}{x^3}$$

Hence, $f'(8) = \dfrac{1}{4\sqrt[3]{8^2}} - \dfrac{24}{8^3} = \dfrac{1}{16} - \dfrac{3}{64} = \dfrac{4}{64} - \dfrac{3}{64} = \dfrac{1}{64}$

BOOST

Grade ⬆⬆⬆⬆

You should be able to do this simple fraction without using a calculator but failing that make sure you know how to do all fraction operations on your calculator.

7.4 Stationary points

A stationary point is a point on a curve where the gradient is zero. A tangent to the curve at a stationary point will have zero gradient and therefore be parallel to the x-axis.

To find the stationary points on a curve you first differentiate the equation of the curve and then substitute the derivative equal to zero. The resulting equation is solved to find the x-coordinate or coordinates of the stationary points.

Maximum and minimum points

Look carefully at the graph drawn here and notice the way the sign of the gradient changes either side of a stationary point for a maximum and minimum.

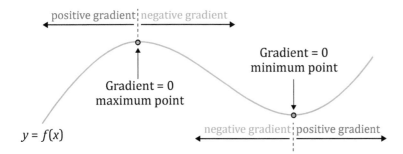

Maximum and minimum points are also called 'turning points' because the gradient changes sign on going from one side of them to the other.

Point of inflection

A point of inflection is a stationary point on a curve (i.e. where the gradient is zero) but where the gradient does not change either side of the stationary point.

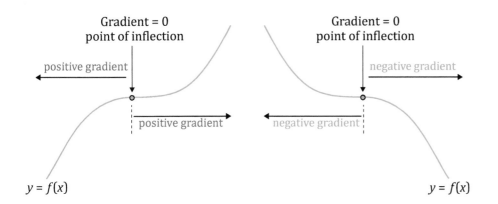

Points of inflection are not turning points as the gradient does not change sign either side of them.

7.5 The second order derivative

In order to find the second derivative $\left(\text{i.e. } \dfrac{d^2y}{dx^2} \text{ or } f''(x)\right)$ you take the first derivative $\left(\text{i.e. } \dfrac{dy}{dx} \text{ or } f'(x)\right)$ and differentiate it again.

The second derivative gives the following information about the stationary points:

- If $\dfrac{d^2y}{dx^2}$ or $f''(x) < 0$ the point is a maximum point.

- If $\dfrac{d^2y}{dx^2}$ or $f''(x) > 0$ the point is a minimum point.

- If $\dfrac{d^2y}{dx^2}$ or $f''(x) = 0$ this gives no further information about the nature of the point and further investigation is necessary.

Remember that second derivatives are used to find the nature of a stationary point (i.e. whether it is a maximum, minimum or point of inflection).

When factorising, care should be taken when extracting a negative number as shown here. A common error is simply to divide through by a common factor before equating to zero, which can result in a change of sign – which in turn could lead to a mis-identification of the nature of any stationary points.

Example

1   The curve C has equation: $y = -2x^3 + 3x^2 + 12x - 5$

Find the coordinates and nature of each of the stationary points of C.   [6]

Answer

1        $y = -2x^3 + 3x^2 + 12x - 5$

$$\frac{dy}{dx} = -6x^2 + 6x + 12 = -6(x^2 - x - 2) = -6(x - 2)(x + 1)$$

At the stationary points, $\frac{dy}{dx} = 0$, so $-6(x - 2)(x + 1) = 0$

Solving gives $x = 2$ or -1

To find the corresponding y values, each of these values is substituted into the equation for the curve.

When $x = 2$, $y = -2(2)^3 + 3(2)^2 + 12(2) - 5 = 15$

When $x = -1$, $y = -2(-1)^3 + 3(-1)^2 + 12(-1) - 5 = -12$

Stationary points are (2, 15) and (−1, −12)

To find the nature of the stationary points, $\frac{dy}{dx}$ is differentiated again.

$$\frac{d^2y}{dx^2} = -12x + 6$$

Each x value is entered in turn to determine whether the second derivative is positive or negative. If negative, the point is a maximum, and, if positive, the point is a minimum.

Hence when $x = 2$,

$$\frac{d^2y}{dx^2} = -12(2) + 6 = -18 < 0 \text{ showing there is a maximum point when } x = 2.$$

When $x = -1$,

$$\frac{d^2y}{dx^2} = -12(-1) + 6 = 18 > 0 \text{ showing there is a minimum point when } x = -1.$$

Hence (2, 15) is a maximum point and (−1, −12) is a minimum point.

7.6 Increasing and decreasing functions

Curves have changing gradients. Depending on the x-coordinate of a point on the curve the gradient, as given by $\frac{dy}{dx}$ or $f'(x)$, can have a positive, negative or zero value.

You may have to show that a particular function is increasing or decreasing at a given point. To do this, you find the gradient by differentiating the equation of the curve and then substituting in the x-coordinate of the given point to see whether the gradient is positive or negative.

If the gradient is positive, then the curve at that point is an increasing function.

If the gradient is negative, then the curve at that point is a decreasing function.

Example

1 A curve C has the equation $y = x^3 - 6x^2 + 2x - 1$

Determine whether y is an increasing or decreasing function at $x = 2$.

> You need to find the gradient and then substitute the x-coordinate into the expression to see whether it is positive or negative.

Answer

1 Differentiating gives $\dfrac{dy}{dx} = 3x^2 - 12x + 2$

When $x = 2$, $\quad \dfrac{dy}{dx} = 3(2)^2 - 12(2) + 2 = 12 - 24 + 2 = -10$

The gradient at $x = 2$ is negative, showing that y is a decreasing function at this point.

Step by STEP

Find the range of values of x for which the function

$$f(x) = x^3 - 5x^2 - 8x + 13$$

is an increasing function.

Steps to take

1 Differentiate to find an expression for the gradient.

2 Need to now find those values of x which will result in the gradient being positive.

3 Form an inequality by letting the derivative be greater than zero.

4 Solve this inequality to find the range of values of x.

Answer

$$f(x) = x^3 - 5x^2 - 8x + 13$$
$$f'(x) = 3x^2 - 10x - 8$$

> Differentiate the function to find $f'(x)$ which gives an expression for the gradient.

For an increasing function, the gradient has to be greater than zero.

Hence $f'(x) > 0$ so $3x^2 - 10x - 8 > 0$

Factorising to find the critical values $(3x + 2)(x - 4) = 0$

Solving gives the critical values of $-\frac{2}{3}$ and 4

The graph of $f'(x)$ is U-shaped, so the sections required for $f'(x) > 0$ are above the x-axis. Hence, for an increasing original function $x < -\frac{2}{3}$ or $x > 4$.

7.7 Simple optimisation problems

Suppose you have the following problem:

You are given a rectangular sheet of metal having dimensions 16 cm by 10 cm.

A square of metal is to be cut out of each corner as shown in the following diagram.

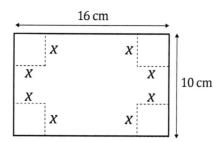

After removing the corners and folding up the flaps an open top box is formed like that shown here:

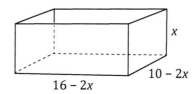

Find the value of x that will make the volume of the box a maximum, and find this maximum volume.

Volume of the box, $V = (16 - 2x)(10 - 2x)(x) = (160 - 52x + 4x^2)(x)$

$$= 160x - 52x^2 + 4x^3$$

Differentiating V with respect to x gives the following:

$$\frac{dV}{dx} = 160 - 104x + 12x^2 = 4(3x^2 - 26x + 40) = 4(3x - 20)(x - 2)$$

For stationary points $\dfrac{dV}{dx} = 0$

$$4(3x - 20)(x - 2) = 0$$

Giving $x = \dfrac{20}{3}$ or 2

$x = \frac{20}{3} = 6\frac{2}{3}$ is an impossible answer as the box is only 10 cm wide so it is not possible to cut two squares of side $6\frac{2}{3}$ cm

Hence $x = 2$ cm

We can check that $x = 2$ is a maximum value by finding the second derivative and then substituting 2 in for x to check that the second derivative is negative.

$$\frac{d^2V}{dx^2} = -104 + 24x$$

When $x = 2$, $\dfrac{d^2V}{dx^2} = -104 + 24(2) = -56$

This is a negative value so the maximum value of V occurs when $x = 2$.

Substituting $x = 2$ into the equation for the volume gives:

$$V = 160(2) - 52(2)^2 + 4(2)^3 = 144$$

Hence, the maximum volume of the box = 144 cm³

Determining whether a stationary point is a point of inflection

The gradient $\left(\text{i.e. } \dfrac{dy}{dx} \text{ or } f'(x)\right)$ at a stationary point is zero.

At a point of inflection there is no change in the sign of the gradient either side of the stationary point.

Example

1 Curve C has the equation $y = x^3 - 6x^2 + 12x - 5$

Find the coordinates of the stationary point on curve C and show that this point is a point of inflection.

. . . .

Answer

1 Differentiating gives $\quad \dfrac{dy}{dx} = 3x^2 - 12x + 12 = 3(x^2 - 4x + 4) = 3(x - 2)^2$

At the stationary point, $\quad \dfrac{dy}{dx} = 0$, so $3(x - 2)^2 = 0$

Solving gives a turning point when $x = 2$

To find the y-coordinate of the stationary point, $x = 2$ is substituted into the equation for the curve.

When $x = 2$, $\quad y = 2^3 - 6(2)^2 + 12(2) - 5 = 3$

Hence the stationary point is at $(2, 3)$.

To show that this is a point of inflection we find the gradient either side of the stationary point and show that the gradient does not change its sign.

$$\dfrac{dy}{dx} = 3(x - 2)^2 \geq 0$$

for all values of x, since any expression squared cannot be negative.

The gradient does not change sign so the stationary point at $x = 2$ is a point of inflection.

> You could also answer this question in a slightly different way. You could find the second derivative and then substitute the x-value of the stationary point into it. If the second derivative is zero, then the point is a point of inflection.

Step by STEP

The diagram below shows a closed box in the form of a cuboid, which is such that the length of its base is twice the width of its base. The volume of the box is $9000\ cm^3$. The total surface area of the box is denoted by $S\ cm^2$.

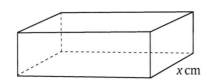

$x\ cm$

(a) Show that $S = 4x^2 + \dfrac{27\,000}{x}$, where $x\ cm$ denotes the width of the base. [3]

(b) Find the minimum value of S, showing that the value you have found is a minimum value. [5]

Steps to take

1   Find the area of the sides. The height and width can be found in terms of x. Need to find the height in terms of x using the equation for the volume of a cuboid.

2   The height in terms of x can now be substituted back into the surface area equation to give the required equation for S.

3   Differentiate the expression for S and then equate it to zero so that the stationary value(s) of x can be found.

4   Determine the nature of the stationary point or points by differentiating again to find the second derivative. Substitute one or each value of x into this to determine the sign and hence the nature of the stationary points (i.e. max or min).

5   Substitute the value of x that gives the minimum back into the equation for S to find the minimum value of S.

. . . .

Answer

(a)   Length $= 2x$ and let height $= y$

$$\text{Area of top + bottom} = 2x^2 + 2x^2 = 4x^2$$

$$\text{Area of front + back} = 2xy + 2xy = 4xy$$

$$\text{Area of two sides} = 2xy$$

$$\text{Total surface area, } S = 6xy + 4x^2$$

$$\text{Volume} = 2x \times x \times y = 2x^2y$$

$$9000 = 2x^2y$$

Hence
$$y = \frac{4500}{x^2}$$

Substituting this expression for y into the expression for the area, we obtain

$$S = 6x \times \frac{4500}{x^2} + 4x^2$$

$$= 4x^2 + \frac{27\,000}{x}$$

(b) $S = 4x^2 + 27\,000\,x^{-1}$

> To differentiate this, you need to first change it to index form.

$$\frac{\mathrm{d}S}{\mathrm{d}x} = 8x - 27\,000\,x^{-2}$$

$$\frac{\mathrm{d}S}{\mathrm{d}x} = 8x - \frac{27\,000}{x^2}$$

At the max/min $\dfrac{\mathrm{d}S}{\mathrm{d}x} = 0$ so $8x - \dfrac{27\,000}{x^2} = 0$

Hence $x^3 = \dfrac{27\,000}{8}$ so $x = 15$

Min/max value of S is $\quad S = 4x^2 + \dfrac{27\,000}{x} = 4(15)^2 + \dfrac{27\,000}{15} = 2700 \text{ cm}^2$

$$\frac{\mathrm{d}^2S}{\mathrm{d}x^2} = 8 + 54\,000x^{-3} = 8 + \frac{54\,000}{x^3}$$

> Notice the square as the denominator. This means whatever value x takes, other than zero, will result in a positive value for the fraction.

When $x = 15$, $\dfrac{\mathrm{d}^2S}{\mathrm{d}x^2} = 8 + \dfrac{54\,000}{x^3}$ = a positive value so the value of S is a minimum value.

7.8 Gradients of tangents and normals, and their equations

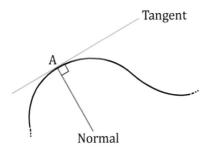

The tangent to a curve and the normal to the curve at the same point are perpendicular to each other.

The gradient of the tangent at point A is the same as the gradient of the curve at point A.

If two lines are perpendicular, the product of their gradients is −1.

> **Important note**
> You may need to look back at Topic 3 on coordinate geometry and straight lines before looking at the rest of this section.

To find the equation of the tangent to a curve at a point P (x, y)

1   Differentiate the equation of the curve to find the gradient.

2   Substitute the x-coordinate of P into $\dfrac{dy}{dx}$ to obtain the gradient of the tangent at P, m.

3   Using the formula for the equation of a straight line: $y - y_1 = m(x - x_1)$, substitute the gradient m, and the coordinates of point P for x_1 and y_1 into the above formula. Rearrange the equation if necessary so that the equation is in the format asked for in the question.

To find the equation of a normal to a curve at a point P (x, y)

1   Differentiate the equation of the curve to find the gradient $\dfrac{dy}{dx}$.

2   Substitute the x-coordinate of P into $\dfrac{dy}{dx}$ to obtain the gradient of the tangent at P, m_1.

3   Find the gradient of the normal using $m_1 m_2 = -1$, i.e. $m_2 = -\dfrac{1}{m_1}$.

4   Using the formula for the equation of a straight line, $y - y_1 = m(x - x_1)$, substitute the gradient m_2, and the coordinates of point P for x_1 and y_1 into the above formula. Rearrange the equation if necessary so that the equation is in the format asked for in the question.

Examples

1   The curve C has equation $y = \dfrac{6}{x^2} + \dfrac{7x}{4} - 2$. The point P has coordinates (2, 3) and lies on C.

Find the equation of the **normal** to C at P. [6]

Reducing the index –2, by 1 gives –3

This is the numerical value for the gradient of the tangent. This is then used to determine the gradient of the normal which is at right angles to the tangent.

Answer

1    $y = \dfrac{6}{x^2} + \dfrac{7x}{4} - 2.$

Substituting this equation in index form gives

$$y = 6x^{-2} + \frac{7x}{4} - 2$$

$$\frac{dy}{dx} = -12x^{-3} + \frac{7}{4}$$

$$\frac{dy}{dx} = -\frac{12}{x^3} + \frac{7}{4}$$

When $x = 2$ $\dfrac{dy}{dx} = -\dfrac{12}{8} + \dfrac{7}{4} = \dfrac{1}{4}$

To find the gradient of the normal we use $m_1 m_2 = -1$

So, $\left(\dfrac{1}{4}\right) m_2 = -1$ (m_2 is the gradient of the normal)

Giving gradient of the normal, $m_2 = -4$

The equation of a straight line having gradient m and passing through the point (x_1, y_1) is given by:

$$y - y_1 = m(x - x_1) \qquad \text{where } m = -4 \text{ and } (x_1, y_1) = (2, 3)$$

So $y - 3 = -4(x - 2)$

$$y = -4x + 11$$

---

2   The curve C has equation $y = \dfrac{1}{2}x^3 - 6x + 3$

Find the coordinates and the nature of each of the stationary points of C.   [6]

Answer

2    $y = \dfrac{1}{2}x^3 - 6x + 3$

Differentiating gives $\dfrac{dy}{dx} = \dfrac{3}{2}x^2 - 6$

At the stationary point, $\dfrac{dy}{dx} = 0$

Hence $\dfrac{3}{2}x^2 - 6 = 0$

Giving $x^2 = 4$

$x = \pm 2$ (Note that you must include both solutions for $\sqrt{4}$)

Finding the second derivative:

$$\frac{d^2y}{dx^2} = 3x$$

The second derivative is found by differentiating the first derivative.

When $x = -2$, $\dfrac{d^2y}{dx^2} = 3 \times (-2) = -6$. This is a negative value, indicating that the stationary point at $x = -2$ is a maximum.

When $x = 2$, $\dfrac{d^2y}{dx^2} = 3 \times 2 = 6$. This is a positive value, indicating that the stationary point at $x = 2$ is a minimum.

To find the y-coordinate for each x-coordinate of the stationary points involves substituting the x-coordinate into the equation of the curve.

When $x = 2$, $\qquad\qquad y = \dfrac{1}{2} \times 8 - 6 \times 2 + 3 = -5$

When $x = -2$, $\qquad\qquad y = \dfrac{1}{2} \times (-8) - 6 \times (-2) + 3 = 11$

Hence there is a maximum point at $(-2, 11)$ and a minimum point at $(2, -5)$.

---

3 The curve C has equation $y = x^2 - 8x + 10$

(a) The point P has coordinates $(3, -5)$ and lies on C. Find the equation of the normal to C at P. [5]

(b) The point Q lies on C and is such that the tangent to C at Q has equation $y = 4x + c$, where c is a constant. Find the coordinates of Q and the value of c. [4]

. . . .

Answer

3 (a) Differentiating the equation of the curve to find the gradient gives
$$\frac{dy}{dx} = 2x - 8$$
At P $(3, -5)$ the gradient is found by substituting $x = 3$ into the expression for $\dfrac{dy}{dx}$

Hence $\qquad\qquad \dfrac{dy}{dx} = 2(3) - 8 = -2$

The tangent and normal are perpendicular to each other, so

$(-2)m = -1$, giving $m = \dfrac{1}{2}$.

The equation of the normal having gradient $\dfrac{1}{2}$ and passing through $(3, -5)$ is

$$y - (-5) = \frac{1}{2}\left(x - 3\right)$$
$$y + 5 = \frac{1}{2}\left(x - 3\right)$$
$$2y + 10 = x - 3$$
$$x - 2y - 13 = 0$$

> The product of the gradients of perpendicular lines is -1. (i.e. $m_1 m_2 = -1$)

(b) The tangent has equation $y = 4x + c$

Gradient of the tangent $= 4$

Gradient of curve $= \dfrac{dy}{dx} = 2x - 8$

Hence $2x - 8 = 4$, giving $x = 6$.

To find the y-coordinate of Q, substitute $x = 6$ into the equation of the curve.

$$y = x^2 - 8x + 10$$
$$y = 6^2 - 8(6) + 10 = 36 - 48 + 10 = -2$$

So the coordinates of Q are $(6, -2)$

As the point Q lies on the tangent, the coordinates of Q must satisfy the equation of the tangent. Substituting $x = 6$ and $y = -2$ into the equation of the tangent gives:

$$y = 4x + c$$

$$-2 = 4(6) + c$$

Giving $\qquad c = -26$

7.9 Simple curve sketching

To sketch a curve when you are given its equation, you need to determine the following:

- The coordinates of the stationary points on the curve and their nature (i.e. whether they are maxima, minima or points of inflection).
- The coordinates of the points where the curve intersects (i.e. crosses) the x-axis.
- The coordinates of the point(s) where the curve intersects the y-axis.

You have already come across how to find the coordinates and the nature of the stationary points.

To find where a curve cuts the x-axis you substitute $y = 0$ and then solve the resulting equation in x.

To find where a curve cuts the y-axis you substitute $x = 0$ into the equation of the curve.

Once you have all these coordinates you can plot them on a suitable set of axes. The graph does not have to be drawn accurately as it is only a sketch but you have to include the coordinates of the points and make sure that the curve is drawn smoothly.

The following example shows all these techniques.

Examples

1   A curve C has the equation $y = x^2 - 2x - 3$

    (a) Find the coordinates and nature of the stationary point of C.   [4]

    (b) Sketch the curve $y = x^2 - 2x - 3$.   [4]

. . . .

Answer

1   (a) $y = x^2 - 2x - 3$

        $\dfrac{dy}{dx} = 2x - 2$

> You could be asked to answer a question such as this using a different method, e.g. completing the square. For information on completing the square, look back at Topic 2.

        At the stationary point$\qquad \dfrac{dy}{dx} = 0$

        Hence$\qquad\qquad\qquad 2x - 2 = 0$

        Solving for x gives$\qquad\quad x = 1$

        Substituting $x = 1$ into the equation of the curve to find the corresponding y-coordinate gives

$$y = 1^2 - 2(1) - 3 = -4$$

Hence the coordinates of the stationary point are $(1, -4)$

Differentiating again to find the nature of the stationary point:

$$\frac{d^2y}{dx^2} = 2$$

The second derivative is positive showing that $(1, -4)$ is a minimum point.

(b) To determine where the curve cuts the x-axis, substitute $y = 0$ into the equation of the curve.

$$0 = x^2 - 2x - 3$$

Factorising gives $(x - 3)(x + 1) = 0$

Solving gives $x = 3$ or -1

To determine where the curve cuts the y-axis, substitute $x = 0$ into the equation of the curve.

$$y = (0)^2 - 2(0) - 3 = -3$$

You now need to draw a set of axes making sure that all the important points you have just found can be shown.

Substitute numbers on each axis where the curve cuts and make sure you mark on the curve the coordinates of the stationary point.

Remember to mark both axes and to write the equation next to the curve.

The curve can now be sketched like this:

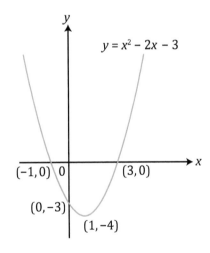

BOOST

Grade ⬆⬆⬆⬆

Not adding the points and not properly labelling the axes frequently cause lost marks in these sorts of questions.

---

2   The curve C has equation

$$y = x^3 - 3x^2 - 9x + 14$$

Sketch C, indicating the coordinates of each of the stationary points of C.

. . . .

Answer

2                    $y = x^3 - 3x^2 - 9x + 14$

$$\frac{dy}{dx} = 3x^2 - 6x - 9$$

At the stationary points, $\dfrac{dy}{dx} = 0$ so $\qquad 3x^2 - 6x - 9 = 0$

Dividing through by 3 gives $\qquad\qquad x^2 - 2x - 3 = 0$

Factorising we obtain $(x - 3)(x + 1)$, giving $x = 3$ or -1

Finding the corresponding y-values: when

$$x = 3, y = 3^3 - 3(3)^2 - 9(3) + 14 = -13$$
$$x = -1, y = (-1)^3 - 3(-1)^2 - 9(-1) + 14 = 19$$

Hence stationary points are $(3, -13)$ and $(-1, 19)$.

$$\frac{d^2y}{dx^2} = 6x - 6$$

When $x = 3$, $\qquad \dfrac{d^2y}{dx^2} = 6(3) - 6 = 12$

(positive indicating stationary point at $x = 3$ is minimum)

$x = -1$, $\qquad \dfrac{d^2y}{dx^2} = 6(-1) - 6 = -12$

(negative indicating stationary point at $x = -1$ is maximum)

So adding the Max at $(-1, 19)$ and Min at $(3, -13)$ and drawing the graph we obtain

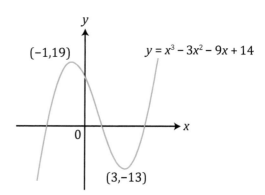

BOOST

Grade ⇧⇧⇧⇧

You must remember to mark the actual coordinates of the stationary values on the graph.

Active Learning

Produce a summary of the steps you need to take if you are asked to find the maximum and minimum values of a function. Write these on a postcard-sized piece of paper or card.

Take a photo of it with your phone and use it for your revision.

Test yourself

1   (a) Given that $y = 4x^2 + 2x - 1$ find $\dfrac{dy}{dx}$ from first principles.

   (b) Given that $y = \dfrac{8}{x^2} + 5\sqrt{x} + 1$, find the gradient of the curve where $x = 1$.

2   The curve C has the following equation:

   $$y = 4\sqrt{x} + \dfrac{32}{x} - 3$$

   (a) Find the values of $\dfrac{dy}{dx}$ when $x = 4$.

   (b) Find the equation of the normal to C at the point where $x = 4$ and where the gradient of the tangent is -1.

3   The curve C has equation: $y = \dfrac{2}{3}x^3 + \dfrac{1}{2}x^2 - 6x$
   Find the coordinates of the stationary points of C and determine the nature of these points.

4   A function is given by $f(x) = \sqrt[3]{x} + 2x + 5$
   Determine whether $f(x)$ is an increasing or decreasing function when $x = 1$.

5   A sheep pen is to be created using 100 m of fencing.
   (a) Letting x be the length of this pen, find the length and width of the pen that would make the area of the pen a maximum.
   (b) Find the area of the resulting pen.

6   Given that $y = x^3$, find $\dfrac{dy}{dx}$ from first principles.   [6]

7   Given that $y = 2x^2 - 7x + 5$, show from first principles that $\dfrac{dy}{dx} = 4x - 7$.   [5]

8   Differentiate $6x^{\frac{2}{3}} - \dfrac{3}{x^3}$ with respect to x.   [2]

9   The curve C has equation $y = x^2 - 8x + 6$. The point A has coordinates $(1, 2)$.
   (a) Find the equation of the tangent to C at A.   [4]
   (b) Find the equation of the normal to C at the point A.   [2]

10 The curve C has equation: $y = x^3 - 3x^2 + 3x + 5$
   (a) Show that C has only one stationary point. Find the coordinates of this point.   [4]
   (b) Verify that this stationary point is a point of inflection.   [2]

11 (a) Given that $y = 3x^2 - 7x - 5$, find $\dfrac{dy}{dx}$ from first principles.   [5]

   (b) Given that $y = ax^{\frac{5}{2}}$, and $\dfrac{dy}{dx} = -2$ when $x = 4$, find the value of the constant a.   [3]

Summary

Check you know the following facts:

Differentiating

To differentiate terms of a polynomial expression:

- Multiply by the index and then reduce the index by one.
- If $y = kx^n$ then the derivative $\frac{dy}{dx} = nkx^{n-1}$.

Increasing or decreasing functions

To find whether a curve or function is increasing or decreasing at a certain point:

- Differentiate the equation of the curve or the function.
- Substitute the value of x at the given point into the expression for the derivative to see whether the gradient is positive or negative. If the value is positive, the function is increasing at the given point, and if the value is negative, the function is decreasing.

Finding a stationary point

- Substitute $\frac{dy}{dx} = 0$ and solve the resulting equation to find the value or values of x at stationary points.
- Substitute the value or values of x into the equation of the curve to find the corresponding y-coordinate(s).

Finding whether a stationary point is a maximum or minimum

- Differentiate the first derivative $\left(\text{i.e. } \frac{dy}{dx}\right)$ to find the second derivative $\left(\text{i.e. } \frac{d^2y}{dx^2}\right)$.
- Substitute the x-coordinate of the stationary point into the expression for $\frac{d^2y}{dx^2}$.
- If the resulting value is negative then the stationary point is a maximum point and if the resulting value is positive, then the stationary point is a minimum point. If $\frac{d^2y}{dx^2} = 0$, then the result is inconclusive and further investigation is required.

Determining whether a stationary point is a point of inflection

- Substitute the x-coordinate of a point either side of the stationary point into the expression for $\frac{dy}{dx}$ and if the gradient has the same sign then the stationary point is a point of inflection.

Curve sketching

- Find the points of intersection with the x and y axes by substituting $y = 0$ and $x = 0$ in turn and then solving the resulting equations.
- Find the stationary points and their nature (i.e. maximum, minimum, point of inflection).
- Plot the above on a set of axes.
- It is important to note that maxima and minima are local maximum and minimum points only, and not necessarily the maximum or minimum values for a function. For example, the graph of a cubic equation shows this clearly.

8 Integration

Introduction

Integration is the reverse process to differentiation. So, if you have the derivative of a function and want to get back to the original function, you would integrate.

Integration is also used to find the area under a curve.

8.1 Indefinite integration as the reverse process of differentiation

Integration is the opposite of differentiation.

For example: If $y = x^2 + 3x + 5$, then $\frac{dy}{dx} = 2x + 3$, so $\int(2x + 3)dx = x^2 + 3x + c$. Notice why a constant of integration called c is needed. When differentiating any constant terms disappear so when integrating it is impossible to know what the value of the constant term should be. Hence a constant c is added. Later you will see how the value of this constant can be found in certain circumstances.

So to integrate x^n you increase the index by one and then divide by the new index. It is important to note that this works for all values of n provided $n \neq -1$. For indefinite integration you must always remember to include the constant of integration, called c.

This can be expressed in the following way:

$$\int x^n \, dx = \frac{x^{n+1}}{n+1} + c \qquad \text{(provided } n \neq -1\text{)}$$

You will see how this works by looking at the following examples:

1   $\int x^3 \, dx = \frac{x^4}{4} + c$

2   $\int 2x \, dx = \frac{2x^2}{2} + c = x^2 + c$

3   $\int 4 \, dx = 4x + c$

The following expression is integrated as follows:

$$\int(x^3 + 4x^2 - x + 2)dx = \frac{x^4}{4} + \frac{4x^3}{3} - \frac{x^2}{2} + 2x + c$$

> This is called an indefinite integral because the result is not definite. You must remember to add a constant of integration c.

Fractional powers can be integrated in the following way

$$\int\left(x^{\frac{1}{2}} + x^{\frac{1}{3}}\right)dx = \frac{x^{\frac{3}{2}}}{\frac{3}{2}} + \frac{x^{\frac{4}{3}}}{\frac{4}{3}} + c = \frac{2}{3}x^{\frac{3}{2}} + \frac{3}{4}x^{\frac{4}{3}} + c$$

> Remember that when you divide by a fraction you must invert the fraction and then multiply to give the answer.

If you have an integral with roots or reciprocals, you must change them to indices before integrating as the following example shows

$$= \int\left(\sqrt{x} + \frac{1}{x^2}\right) = \int\left(x^{\frac{1}{2}} + x^{-2}\right)dx = \frac{x^{\frac{3}{2}}}{\frac{3}{2}} + \frac{x^{-1}}{-1} + c = \frac{2}{3}x^{\frac{3}{2}} - x^{-1} + c$$

> There are lots of conversions to indices and back again in this Topic. If you are unsure about indices, look back at Topic 1.

Examples

1   Find $\int(x^3 + 3x^2 - 2x + 1)dx$

. . . .

Answer

1   $\int(x^3 + 3x^2 - 2x + 1)dx = \frac{x^4}{4} + \frac{3x^3}{3} - \frac{2x^2}{2} + x + c$

$$= \frac{x^4}{4} + x^3 - x^2 + x + c$$

> Note the reciprocal and the root in this integral. Both must be changed to index form so they can be integrated.

2   Find $\int\left(3x^2 + \frac{1}{x^2} + \sqrt{x}\right)dx$

Answer

2 $\displaystyle\int\left(3x^2 + \frac{1}{x^2} + \sqrt{x}\right)dx = \int\left(3x^2 + x^{-2} + x^{\frac{1}{2}}\right)dx$

$$= \frac{3x^3}{3} + \frac{x^{-1}}{-1} + \frac{x^{\frac{3}{2}}}{\frac{3}{2}} + c$$

$$= x^3 - x^{-1} + \frac{2}{3}x^{\frac{3}{2}} + c$$

Note that when you divide by a fraction, you invert the fraction and then multiply by it. Hence:

$$\frac{x^{\frac{3}{2}}}{\frac{3}{2}} = \frac{2}{3}x^{\frac{3}{2}}$$

Be careful with the signs when dealing with negative indices.

---

3 Find $\displaystyle\int\left(4x^3 - \frac{2}{\sqrt{x}}\right)dx$

Answer

3 $\displaystyle\int\left(4x^3 - \frac{2}{\sqrt{x}}\right)dx = \int\left(4x^3 - 2x^{-\frac{1}{2}}\right)dx$

$$= \frac{4x^4}{4} - \frac{2x^{\frac{1}{2}}}{\frac{1}{2}} + c$$

$$= x^4 - 4x^{\frac{1}{2}} + c$$

Note that $\dfrac{1}{\sqrt{x}} = x^{-\frac{1}{2}}$

Remember to include c, the constant of integration.

---

4 Find $\displaystyle\int\left(\sqrt[3]{x} - \frac{2}{x^2}\right)dx$

Answer

4 $\displaystyle\int\left(\sqrt[3]{x} - \frac{2}{x^2}\right)dx = \int\left(x^{\frac{1}{3}} - 2x^{-2}\right)dx$

$$= \frac{x^{\frac{4}{3}}}{\frac{4}{3}} - \frac{2x^{-1}}{-1} + c$$

$$= \frac{3}{4}x^{\frac{4}{3}} + 2x^{-1} + c$$

Be careful when increasing a negative power by 1. Check that the answer you get is greater than before.

---

5 Find $\displaystyle\int\left(\frac{4}{x^3} - 3x^{\frac{3}{4}}\right)dx$

Answer

5 $\displaystyle\int\left(\frac{4}{x^3} - 3x^{\frac{3}{4}}\right)dx = \int\left(4x^{-3} - 3x^{\frac{3}{4}}\right)dx$

$$= \frac{4x^{-2}}{-2} - \frac{3x^{\frac{7}{4}}}{\frac{7}{4}} + c$$

$$= -2x^{-2} - \frac{12x^{\frac{7}{4}}}{7} + c$$

---

6 Find $\displaystyle\int\left(\frac{3}{\sqrt[4]{x}} - 9x^{\frac{5}{2}}\right)dx$

Answer

6 $\displaystyle\int\left(\frac{3}{\sqrt[4]{x}} - 9x^{\frac{5}{2}}\right)dx = \int\left(3x^{-\frac{1}{4}} - 9x^{\frac{5}{2}}\right)dx$

$$= \frac{3x^{\frac{3}{4}}}{\frac{3}{4}} - \frac{9x^{\frac{7}{2}}}{\frac{7}{2}} + c$$

$$= 4x^{\frac{3}{4}} - \frac{18x^{\frac{7}{2}}}{7} + c$$

8.2 Interpretation of the definite integral as the area under a curve

Integrals in the form $\int_a^b y\,dx$ are called definite integrals because the result will be a definite answer, usually a number, with no constant of integration.

The definite integral is found by substituting the limits into the result of the integration, and subtracting the value corresponding to the lower limit from the value corresponding to the upper limit.

A definite integral is positive for areas above the x-axis and negative for areas below the x-axis. A final area must always be given as a positive value.

Definite integrals represent the area under the curve $y = f(x)$ between the two x-values $x = a$ and $x = b$.

The vertical lines drawn from $x = a$ and $x = b$ which meet the curve are called ordinates.

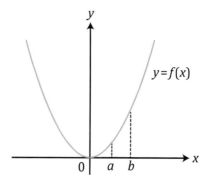

Examples

1   Find $\int_1^2 (3x^2 - x + 4)\,dx$

Answer

1   $\int_1^2 (3x^2 - x + 4)\,dx = \left[\frac{3x^3}{3} - \frac{x^2}{2} + 4x\right]_1^2$

Once you have integrated, put square brackets around the result and write the limits as shown here.

$$= \left[x^3 - \frac{x^2}{2} + 4x\right]_1^2$$

Two pairs of brackets are used. The first contains the top limit substituted in for x. The second contains the bottom limit substituted in for x. The contents of the second bracket are subtracted from the contents of the first.

$$= \left[\left(2^3 - \frac{2^2}{2} + 4(2)\right) - \left(1^3 - \frac{1^2}{2} + 4(1)\right)\right]$$

$$= \left[\left(8 - 2 + 8\right) - \left(1 - \frac{1}{2} + 4\right)\right]$$

$$= 14 - 4\tfrac{1}{2} = 9\tfrac{1}{2} \text{ square units}$$

2 Find $\int_0^3 \left(\frac{4x^3}{3} - \frac{x^2}{2}\right)dx$

Answer

2
$$\int_0^3 \left(\frac{4x^3}{3} - \frac{x^2}{2}\right)dx = \left[\frac{4x^4}{12} - \frac{x^3}{2 \times 3}\right]_0^3$$

$$= \left[\frac{x^4}{3} - \frac{x^3}{6}\right]_0^3$$

$$= \left[\left(\frac{3^4}{3} - \frac{3^3}{6}\right) - (0)\right]$$

$$= \left[\left(27 - 4.5\right) - (0)\right]$$

$$= 22.5$$

> If one of the limits is zero then if all the terms in the bracket to which this limit refers contain x, then you can simply replace the whole bracket with a zero.

3 (a) Find $\int \left(\frac{2}{\sqrt{x}} - x^3 + \frac{2}{x^2}\right)dx$

(b)

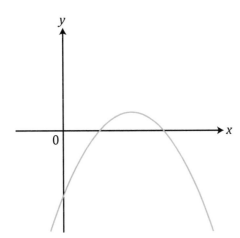

The diagram shows a sketch of the curve $y = (1 - x)(x - 3)$.

(i) Find the coordinates of the points of intersection of the curve and the x-axis.

(ii) Find the area of the shaded region.

Answer

3 (a) $\int \left(\frac{2}{\sqrt{x}} - x^3 + \frac{2}{x^2}\right)dx = \int \left(2x^{-\frac{1}{2}} - x^3 + 2x^{-2}\right)dx$

$$= \frac{2x^{\frac{1}{2}}}{\frac{1}{2}} - \frac{x^4}{4} + \frac{2x^{-1}}{-1} + c$$

$$= 4x^{\frac{1}{2}} - \frac{x^4}{4} - 2x^{-1} + c$$

> Change to indices so that the expression can be integrated.

BOOST

Grade ⇧⇧⇧⇧

Forgetting to include the constant of integration frequently costs students marks.

(b) (i) When $y = 0$, $(1 - x)(x - 3) = 0$

Giving $x = 1$ or $x = 3$

> Put the equation of the curve equal to 0 to find the points of intersection with the x-axis.

The coordinates of the points of intersection with the x-axis are $(1, 0)$ and $(3, 0)$.

(ii) Shaded area $= \int_1^3 y\,dx = \int_1^3 (1-x)(x-3)\,dx$

$$= \int_1^3 (-3 + 4x - x^2)\,dx$$

$$= \left[-3x + 2x^2 - \frac{x^3}{3} \right]_1^3$$

$$= \left[\left(-9 + 18 - 9 \right) - \left(-3 + 2 - \tfrac{1}{3} \right) \right]$$

$$= 0 - \left(-\tfrac{4}{3} \right)$$

$$= \tfrac{4}{3} \text{ square units}$$

---

4 (a) Draw a sketch of the curve $y = x^3$.

(b) Find $\int_{-2}^0 x^3 \, dx$

(c) Find the area enclosed by the curve and the ordinates $x = -2$ and $x = 2$.

. . . .

Answer

4 (a)

> Note the symmetry of the graph.

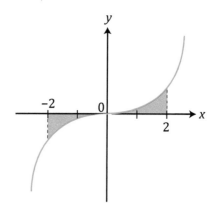

> Note that the area is negative as areas below the x-axis are always negative.

(b) $\int_{-2}^0 x^3 \, dx = \left[\frac{x^4}{4} \right]_{-2}^0 = \left[\left(\frac{(0)^4}{4} \right) - \left(\frac{(-2)^4}{4} \right) \right] = -4$

(c) You could integrate between 0 and 2 and then add the result to that obtained in part (a) but this is not necessary.

Owing to the symmetry, the area of the graph is double the answer obtained in part (a) but we remove the negative sign.

Hence, required area = 8

> Looking at the sketch in part (a) and marking on it the two areas, you can see that one of the areas is below the x-axis.
>
> This will give a negative answer. The positive area above the x-axis is equal in magnitude so if we integrate between the limits −2 and 2 it will give the area as zero.

8.3 Evaluation of definite integrals

Definite integrals have a numerical value which is obtained when the two limits are substituted for x into the result of the integration as the following examples show.

Example

1 Find $\int_0^1 \left(\dfrac{2}{\sqrt{x}} + 7x^{\frac{4}{3}} \right) dx$

Answer

1 $\int_0^1 \left(\dfrac{2}{\sqrt{x}} + 7x^{\frac{4}{3}} \right) dx = \int_0^1 \left(2x^{-\frac{1}{2}} - 7x^{\frac{4}{3}} \right) dx$

$= \left[\dfrac{2x^{\frac{1}{2}}}{\frac{1}{2}} - \dfrac{7x^{\frac{7}{3}}}{\frac{7}{3}} \right]_0^1$

$= \left[4x^{\frac{1}{2}} - 3x^{\frac{7}{3}} \right]_0^1$

$= \left[\left(4(1)^{\frac{1}{2}} - 3(1)^{\frac{7}{3}} \right) - \left(4(0)^{\frac{1}{2}} - 3(0)^{\frac{7}{3}} \right) \right]$

$= 1$

> Make sure all the terms are in index form ready to be integrated.

8.4 Finding the area bound by a straight line and a curve

The area between a curve and a straight line

To find the area bound between a curve and a straight line, first find the points of intersection if they are not already given. Then integrate the curve between the two x-coordinates to find the area under the curve. Then find the area of the remaining shape, usually a triangle or a trapezium and then either add or subtract the two areas to find the required shaded area. The following diagrams show how the area is found in different situations.

In this situation, the shaded region in the first diagram is below the straight line and above the curve. To obtain the required area we subtract the area under the curve from the area of the triangle.

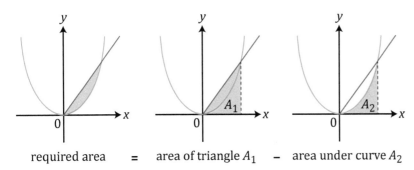

required area = area of triangle A_1 − area under curve A_2

> Simply look at the graphs and think logically about the area you are finding. You can then make the decision about the shapes you need to add or subtract to give the required area.

In this situation the shaded region in the first diagram below is above the straight line and below the curve. To obtain the required area we subtract the area of the triangle from the area under the curve.

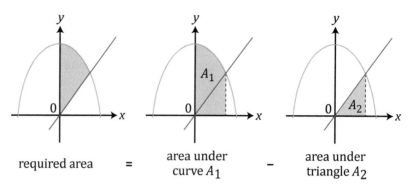

required area	=	area under curve A_1

minus

area under triangle A_2

Suppose we are required to find the shaded area in the graph below:

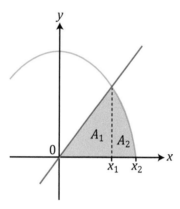

You would need to first find the coordinates of the points of intersection between the straight line and the curve by solving their two equations simultaneously.

The area of the triangle A_1 can be found using the formula

$$A_1 = \tfrac{1}{2} \times \text{base} \times \text{height.}$$

The area under the curve, A_2, between the points x_1 and x_2 can be found by integrating the equation of the curve between the limits x_1 and x_2.

The required area is then obtained by adding the two areas A_1 and A_2.

Hence required area = $A_1 + A_2$.

It is always advisable to draw diagrams showing the areas representing the parts you are going to use to find the required area.

Example

1   The graph below shows the curve with equation $y = x^2$ and the straight line with equation $y = 2x$. They both intersect at the origin and point A.

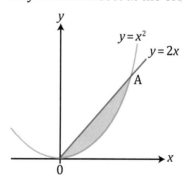

(a)   Find the coordinates of point A.

(b)   Calculate the area of the shaded region.

Answer

1 (a) Equating the y-values of the curve and the straight line we obtain

$$x^2 = 2x$$

$$x^2 - 2x = 0$$

$$x(x - 2) = 0$$

Solving gives $x = 0$ or $x = 2$.

When $x = 2, y = 2(2) = 4$.

Hence coordinates of A are $(2, 4)$.

> Do not divide both sides by x as you will lose one of the solutions if you do this.

(b)

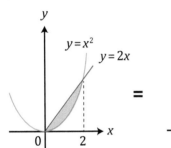 = 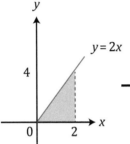 −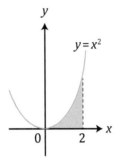

| required area | = | area of triangle between $x = 0$ and $x = 2$ | − | area under curve between $x = 0$ and $x = 2$ |

Area of triangle $= \frac{1}{2}$ base × height

$$= \frac{1}{2} \times 2 \times 4$$

$$= 4$$

Area under curve $= \int_0^2 x^2 dx$

$$= \left[\frac{x^3}{3}\right]_0^2$$

$$= \frac{8}{3}$$

Hence required area $= 4 - \frac{8}{3} = \frac{4}{3}$ or $1\frac{1}{3}$

Step by STEP

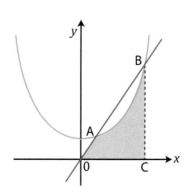

The diagram shows a sketch of the curve $y = x^2 + 3$ and the line $y = 4x$. The curve and the line intersect at the points A and B. The line BC is parallel to the y-axis.

Find the area of the shaded region.

Steps to take

1   First find the coordinates of points A and B the points of intersection of the line and the curve by solving the equations of each simultaneously.

2   Find the area under the curve between the x-coordinate of point A and the x-coordinate of point B by integrating the equation of the curve and putting these limits in.

3   Find the area of the small right-angled triangle formed when a perpendicular is dropped from the point A to the x-axis.

4   Add the area of the small triangle to that found in step 2 to give the area shaded.

· ·

Answer

Equating the equations to find the points of intersection, we obtain

$$x^2 + 3 = 4x \qquad \text{so} \quad x^2 - 4x + 3 = 0$$

Factorising, we obtain $(x - 3)(x - 1) = 0$ giving $x = 1$ or 3

Substituting these values into the equation of the line (as it is simpler) we obtain $x = 1, y = 4$ and $x = 3, y = 12$, so A is the point (1, 4) and B is the point (3, 12).

$$\text{Area under the curve between A and B} = \int_1^3 y\,dx$$

$$= \int_1^3 (x^2 + 3)\,dx$$

$$= \left[\frac{x^3}{3} + 3x\right]_1^3$$

$$= \left[(9 + 9) - \left(\frac{1}{3} + 3\right)\right]$$

$$= \left[18 - 3\frac{1}{3}\right]$$

$$= 14\frac{2}{3}$$

$$\text{Area of small triangle} = \frac{1}{2} \times \text{base} \times \text{height} = \frac{1}{2} \times 1 \times 4 = 2$$

$$\text{Shaded area} = 14\frac{2}{3} + 2 = 16\frac{2}{3} \text{ square units}$$

Example

1 (a) Find $\int\left(5\sqrt{x} - \dfrac{4}{x^{\frac{2}{3}}}\right)dx$ [2]

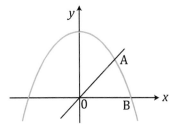

(b) The diagram shows a sketch of the curve $y = 4 - x^2$ and the line $y = 3x$. The curve and the line intersect at the point A in the first quadrant and the curve intersects the positive x-axis at the point B.

(i) Showing your working, find the coordinates of A and the coordinates of B.

(ii) Find the area of the shaded region. [12]

Answer

1 (a) $\int\left(5\sqrt{x} - \dfrac{4}{x^{\frac{2}{3}}}\right)dx = \int\left(5x^{\frac{1}{2}} - 4x^{-\frac{2}{3}}\right)dx$

$= \dfrac{5x^{\frac{3}{2}}}{\frac{3}{2}} - \dfrac{4x^{\frac{1}{3}}}{\frac{1}{3}} + c$

$= \dfrac{2}{3} \times 5x^{\frac{3}{2}} - 3 \times 4x^{\frac{1}{3}} + c$

$= \dfrac{10}{3}x^{\frac{3}{2}} - 12x^{\frac{1}{3}} + c$

> Remember that to integrate you increase the index by one and then divide by the new index.

> Remember when you divide by a fraction, turn the fraction upside down and multiply by the new fraction, so dividing by $\frac{1}{3}$ is the same as multiplying by 3.

(b) Equating the y-values gives

$$3x = 4 - x^2$$

$$x^2 + 3x - 4 = 0$$

Factorising gives $(x - 1)(x + 4) = 0$

Solving gives $x = 1$ or -4

The x-coordinate of A cannot be -4 as A is in the first quadrant.

As $y = 3x$, substituting $x = 1$ into this gives $y = 3(1) = 3$

Hence A is the point $(1, 3)$

For the coordinates of B, substitute $y = 0$ into the equation $y = 4 - x^2$

So, $0 = 4 - x^2$

$x^2 = 4$

$x = \pm 2$

From the question, B has a positive x-value, so $x = 2$.

Hence B has coordinates $(2, 0)$

> As this is an indefinite integral you must remember to include the constant c.

> To find the coordinates of A, solve the equation of the curve simultaneously with that of the straight line.

> Always look at the diagram to check the significance of the points found.

> Remember to include both values of here.

Number or letter the areas on the diagram so they can be easily referred to.

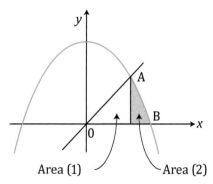

$$\text{Area of (1) = area of a triangle} \quad = \frac{1}{2} \times base \times height$$

$$= \frac{1}{2} \times 1 \times 3 = \frac{3}{2}$$

$$\text{Area (2) under the curve} = \int_1^2 \left(4 - x^2\right)dx$$

$$= \left[\left(4x - \frac{x^3}{3}\right)\right]_1^2$$

$$= \left(8 - \frac{8}{3}\right) - \left(4 - \frac{1}{3}\right)$$

$$= 4 - 2\frac{1}{3}$$

$$= 1\frac{2}{3}$$

$$\text{Total area = area (1) + area (2)} = \frac{3}{2} + 1\frac{2}{3} = 3\frac{1}{6} \text{ square units}$$

Test yourself

1 Find $\displaystyle\int\left(4x^{\frac{1}{3}} - \frac{2}{\sqrt[3]{x}}\right)dx$

2 Find $\displaystyle\int\left(\sqrt[3]{x} - \frac{1}{x^4}\right)dx$

3 Find $\displaystyle\int\left(\frac{4}{x^3} - 6x^{\frac{1}{5}}\right)dx$

4 Find $\displaystyle\int\left(\frac{2}{\sqrt{x}} - x^{\frac{3}{2}}\right)dx$

5 Find $\displaystyle\int_0^4\left(x^{-\frac{1}{2}} + 2x\right)dx$

6 (a) Find $\displaystyle\int\left(\frac{5}{x^3} - 3x^{\frac{1}{4}}\right)dx$

(b)

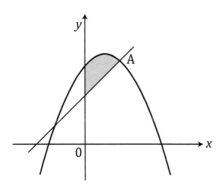

The diagram shows a sketch of the curve $y = 6 + 4x - x^2$ and the line $y = x + 2$. The point of intersection of the curve and the line in the first quadrant is denoted by A.
(i) Find the coordinates of A.
(ii) Find the area of the shaded region. [10]

7 (a) Find $\int \left(\dfrac{3}{x^2} - 2\sqrt{x} \right) dx$

(b)

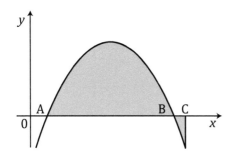

The diagram shows a sketch of the curve $y = 5x - 4 - x^2$.
The curve intersects the x-axis at the points A and B. The point C has coordinates $(5, 0)$.
(i) Find the x-coordinates of the points A and B. [3]
(ii) Find the **total** area of the shaded region. [3]

8

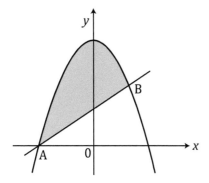

The diagram shows a sketch of the curve $y = 9 - x^2$ and the line $y = x + 3$.
The line and the curve intersect at the points A and B.
Find the coordinates of A and B and area of the shaded region. [7]

Summary

Integration

Indefinite integration is the reverse process of differentiation. When integrating indefinitely you must remember to include the constant of integration.

$$\int x^n \, dx = \frac{x^{n+1}}{n+1} + c \qquad (\text{provided } n \neq -1)$$

A definite integral is positive for areas above the x-axis and negative for areas below the x-axis.

A final area must always be given as a positive value.

9 Vectors

Introduction

Vectors can be used to describe displacements, velocities, accelerations and forces in two or three dimensions. Using vectors simplifies solving real-life problems involving the above.

In this topic we will look at vectors in two dimensions from a pure mathematics perspective. You will come across vectors again when solving problems in mechanics.

9.1 Scalars and vectors

A scalar is a quantity that has size only. It is simply represented by a number and has no direction. Distance, speed and time are scalar quantities.

A vector is a quantity such as force, velocity and displacement (i.e. distance in a specific direction) which has both a size (i.e. magnitude) and a direction. A vector can be represented by a line whose length represents the size and a direction in which the line is pointing.

Note that we have defined a vector in terms of magnitude and direction, but have not been specific about the position of the vector in space. Such vectors are called free vectors.

All the lines below which are parallel and of equal magnitude represent the vector **a**.

Vectors are especially important as they will crop up again in both of the applied papers.

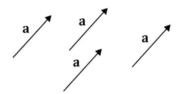

The notation used for vectors

The vector $\overrightarrow{AB}$ can be represented as a single letter in bold (e.g. **r**) but when handwritten it can be written with the letter underlined like r. You may also see the vector $\overrightarrow{AB}$ written without the arrow but in bold, like **AB**.

The length of a vector is written as $|\mathbf{r}|$.

9.2 Vectors in two dimensions

We only have to cover vectors in two dimensions here.

Vectors can be represented in two or three dimensions. Vectors in two dimensions can be drawn in the plane of the paper.

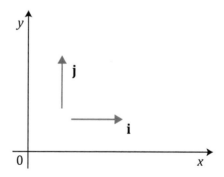

Any vector in two dimensions can be written in terms of the unit vectors **i** and **j** in the following way:

$$\mathbf{r} = a\mathbf{i} + b\mathbf{j} \quad \text{where } a \text{ and } b \text{ are scalars.}$$

For example, here are some vectors expressed in terms of the unit vectors.

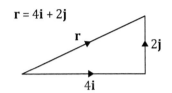

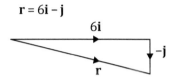

The **i** and **j** vectors make an angle of 90 degrees to each other.

9.3 The magnitude of a vector

The magnitude of a vector

The magnitude of a vector is the length of the vector.

The vector $\mathbf{r} = a\mathbf{i} + b\mathbf{j}$ has magnitude given by $|\mathbf{r}| = \sqrt{a^2 + b^2}$

If a vector is written as $\overrightarrow{AB}$ then its magnitude is written as $|\overrightarrow{AB}|$

Example

1   Find the exact magnitude of the vector $\mathbf{r} = 2\mathbf{i} - 2\mathbf{j}$.

. . . .

Answer

1   $|\mathbf{r}| = \sqrt{2^2 + (-2)^2} = \sqrt{8} = 2\sqrt{2}$

9.4 Algebraic operations of vector addition, subtraction and multiplication by scalars, and their geometrical interpretations

Addition, subtraction and multiplication by a scalar can be performed algebraically as the following example shows.

Example

1   If $\mathbf{a} = \mathbf{i} + 3\mathbf{j}$ and $\mathbf{b} = 3\mathbf{i} - 2\mathbf{j}$, find each of the following:

(a) $\mathbf{a} + \mathbf{b}$

(b) $\mathbf{a} - \mathbf{b}$

(c) $2\mathbf{a}$

(d) $2\mathbf{a} + 3\mathbf{b}$

. . . .

Answer

1   (a) $\mathbf{a} + \mathbf{b} = \mathbf{i} + 3\mathbf{j} + 3\mathbf{i} - 2\mathbf{j} = 4\mathbf{i} + \mathbf{j}$

(b) $\mathbf{a} - \mathbf{b} = \mathbf{i} + 3\mathbf{j} - (3\mathbf{i} - 2\mathbf{j}) = -2\mathbf{i} + 5\mathbf{j}$

(c) $2\mathbf{a} = 2(\mathbf{i} + 3\mathbf{j}) = 2\mathbf{i} + 6\mathbf{j}$

(d) $2\mathbf{a} + 3\mathbf{b} = 2(\mathbf{i} + 3\mathbf{j}) + 3(3\mathbf{i} - 2\mathbf{j}) = 11\mathbf{i}$

Be careful when subtracting two vectors. It is always a good idea to bracket the vector after the minus sign to emphasise it.

The geometrical interpretation of vectors

The geometrical interpretation of the results in 1(a) to (d) of the previous example is as follows:

Addition of vectors, as in 1(a), above

As we are dealing with free vectors, we can move our vector representation.

To add two vectors **a** and **b** as shown, place the starting point of **b** at the finishing point of **a** and complete the triangle. The directed third line represents **a** + **b**.

We could place the starting point of **a** at the finishing point of **b** to form **b** + **a** which equals **a** + **b**. Remember that **a** + **b** and **b** + **a** are free vectors.

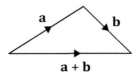

Subtraction of vectors, as in 1(b), above

To draw **a** – **b** , we reverse the direction of **b** to form –**b** and place the starting point of –**b** at the finishing point of **a**.

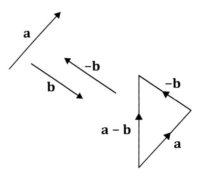

Multiplication of parallel vectors, as in 1(c), above

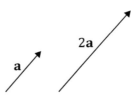

2**a** is a vector parallel to **a** with magnitude equal to twice the magnitude of **a**.

Combination of vectors, as in 1(d), above

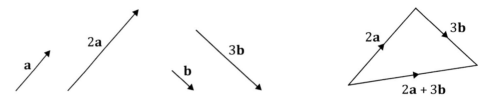

We form 2**a** and 3**b** as shown and use the procedure given in 1(a) to obtain the vector 2**a** + 3**b**

The condition for two vectors to be parallel

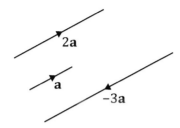

For two vectors **a** and **b** to be parallel one of the vectors must be a multiple of the other, so

$$\mathbf{a} = k\mathbf{b} \quad \text{where } k \text{ is a scalar}$$

For example, if **a** = 3**i** + **j** and **b** = 6**i** + 2**j**, **a** = 2**b** so vectors **a** and **b** are parallel and vector **b** has a magnitude twice that of **a**.

9.5 Position vectors

Position vectors are vectors giving the position of a point, relative to a fixed point (usually the origin). The position vector of a point P relative to the origin O is defined by the vector $\overrightarrow{OP}$. In contrast to a free vector, a position vector is represented by a particular line in space.

If the position vector of A is **a** and the position of vector B is **b** then the following diagram can be drawn showing the two vectors.

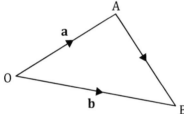

$\overrightarrow{AB} = \overrightarrow{AO} + \overrightarrow{OB} = -\overrightarrow{OA} + \overrightarrow{OB} = -\mathbf{a} + \mathbf{b}$

Hence $\overrightarrow{AB} = \mathbf{b} - \mathbf{a}$

Note that $\overrightarrow{AB}$ can also be written using the alternative notation for vectors as **AB**.

This is an important result and should be remembered.

9.6 Coordinate geometry and vectors

Sometimes you are given the coordinates of the starting and finishing points of a vector (e.g. A $\begin{pmatrix} -2 \\ -6 \end{pmatrix}$ and B $\begin{pmatrix} 3 \\ 6 \end{pmatrix}$). If you wanted to find the vector $\overrightarrow{AB}$ you would subtract x-coordinates from A to B and do the same for the y-coordinates.

So $\overrightarrow{AB} = \begin{pmatrix} 3 - (-2) \\ 6 - (-6) \end{pmatrix} = \begin{pmatrix} 5 \\ 12 \end{pmatrix}$ or $\overrightarrow{AB} = 5\mathbf{i} + 12\mathbf{j}$

If you wanted to find the magnitude of $\overrightarrow{AB}$ you could use the formula for the distance between two points, or employ the formula which uses Pythagoras' theorem when you are given the vector.

Note

Vectors can be written in the **i**, **j** notation, as row vectors or as column vectors.

For example 3**i** − 4**j** is the same as (3, −4) or $\begin{pmatrix} 3 \\ -4 \end{pmatrix}$

You should not use more than one notation in a solution to a question.

9.7 Position vector of a point dividing a line in a given ratio

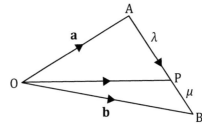

The point P dividing AB in the ratio $\lambda : \mu$ has position vector $\overrightarrow{OP}$

where $\overrightarrow{OP} = \dfrac{\mu \mathbf{a} + \lambda \mathbf{b}}{\lambda + \mu}$

You are required to be able to derive the above formula. It can be derived as follows:

$$\overrightarrow{OP} = \overrightarrow{OA} + \overrightarrow{AP}$$
$$= \mathbf{a} + \frac{\lambda}{\lambda + \mu}\,\overrightarrow{AB}$$
$$= \mathbf{a} + \frac{\lambda}{\lambda + \mu}(\mathbf{b} - \mathbf{a})$$
$$= \frac{(\lambda + \mu)\mathbf{a} + \lambda(\mathbf{b} - \mathbf{a})}{\lambda + \mu}$$
$$= \frac{\mu \mathbf{a} + \lambda \mathbf{b}}{\lambda + \mu}$$

Examples

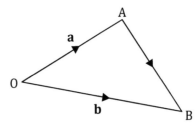

1 The vectors $\mathbf{a} = 3\mathbf{i} + 2\mathbf{j}$ and $\mathbf{b} = 6\mathbf{i} - 3\mathbf{j}$ are the position vectors of A and B respectively.

(a) Find $\overrightarrow{AB}$

(b) Point P divides line AB in the ratio $3 : 1$.

Find the position vector $\overrightarrow{OP}$.

. . . .

Answer

1 (a) $\overrightarrow{AB} = \overrightarrow{AO} + \overrightarrow{OB} = -\overrightarrow{OA} + \overrightarrow{OB} = -\mathbf{a} + \mathbf{b}$

Hence $\overrightarrow{AB} = \mathbf{b} - \mathbf{a} = 6\mathbf{i} - 3\mathbf{j} - (3\mathbf{i} + 2\mathbf{j}) = 3\mathbf{i} - 5\mathbf{j}$

(b)

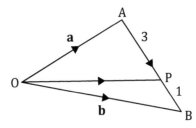

An alternative method to the method shown would be to use the formula
$$\overrightarrow{OP} = \frac{\mu \mathbf{a} + \lambda \mathbf{b}}{\lambda + \mu}$$
which can be obtained from the formula booklet.

Draw a diagram showing point P splitting line AB in the ratio $3 : 1$. Notice that the line is divided into 4 equal parts so AP is three-quarters of the line AB.

$$\overrightarrow{OP} = \overrightarrow{OA} + \overrightarrow{AP}$$

$$= \mathbf{a} + \frac{3}{3+1}\overrightarrow{AB}$$

$$= \mathbf{a} + \frac{3}{4}(3\mathbf{i} - 5\mathbf{j})$$

$$= 3\mathbf{i} + 2\mathbf{j} + \frac{3}{4}(3\mathbf{i} - 5\mathbf{j})$$

$$= 3\mathbf{i} + 2\mathbf{j} + \frac{9}{4}\mathbf{i} - \frac{15}{4}\mathbf{j}$$

$$= \frac{21}{4}\mathbf{i} - \frac{7}{4}\mathbf{j}$$

2 The vectors **a** and **b** are the position vectors of points A and B respectively. P is a point on AB such that AP : PB = 4 : 1. Find the position vector of point P.

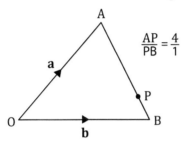

$$\frac{AP}{PB} = \frac{4}{1}$$

If P splits AB in the ratio of 4:1 then the line is being split into 5 parts.

. . . .

Answer

2 Using the formula, the position vector of P, $\overrightarrow{OP} = \dfrac{\mu\mathbf{a} + \lambda\mathbf{b}}{\lambda + \mu}$

As $\lambda = 4$ and $\mu = 1$, $\overrightarrow{OP} = \dfrac{\mathbf{a} + 4\mathbf{b}}{4 + 1} = \dfrac{1}{5}\mathbf{a} + \dfrac{4}{5}\mathbf{b}$

This formula can be obtained from the formula booklet.

3 Point P divides AB in the ratio AP : PB = n : 1 where n is an integer. If the position vector of P is $\frac{1}{10}\mathbf{a} + \frac{9}{10}\mathbf{b}$, find the value of n.

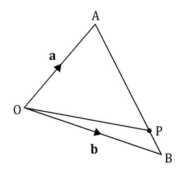

. . . .

Answer

3 Position vector of P, $\overrightarrow{OP} = \dfrac{\mu\mathbf{a} + \lambda\mathbf{b}}{\lambda + \mu}$

But position vector of P, $\overrightarrow{OP} = \dfrac{\mathbf{a} + 9\mathbf{b}}{10}$

Hence $\mu = 1$ and $\lambda = 9$

So line is divided in the ratio $\lambda : \mu = 9 : 1$

So the value of $n = 9$

Proving a point with a certain position vector lies on a line whose vector is known

In the following triangle, P is the point that divides the line AB in the ratio of 1 : 5.

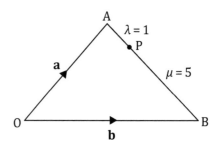

The point P dividing AB in the ratio $\lambda : \mu$ has position vector $\overrightarrow{OP} = \dfrac{\mu\mathbf{a} + \lambda\mathbf{b}}{\lambda + \mu}$

As P divides AB in the ratio of 1 : 5 it has position vector $\overrightarrow{OP} = \dfrac{5\mathbf{a} + \mathbf{b}}{1 + 5} = \dfrac{5}{6}\mathbf{a} + \dfrac{1}{6}\mathbf{b}$

If a different point Q divides AB in the ratio of 2 : 3,

the position vector of Q, $\overrightarrow{OQ} = \dfrac{3\mathbf{a} + 2\mathbf{b}}{2 + 3} = \dfrac{3}{5}\mathbf{a} + \dfrac{2}{5}\mathbf{b}$

If you look at any position vector for a point lying on AB (and therefore dividing AB into a certain ratio) the coefficient of **b** is 1 − coefficient of **a**.

So the point with position vector $\frac{5}{9}\mathbf{a} + \frac{4}{9}\mathbf{b}$ would lie on line AB but the point with position vector $\frac{2}{3}\mathbf{a} + \frac{1}{5}\mathbf{b}$ would not.

9.8 Using vectors to solve problems in pure mathematics

Vectors are used in both pure mathematics and mechanics. You will come across vectors being used in mechanics later. In this section you will learn about how vectors can be used to help solve problems in pure mathematics. For example, you can prove that two lines are parallel or whether a particular point lies on a line or not. You can also find angles between vectors. All of the following problems use the theory you learned in the previous sections.

You will come across vectors when you start the applied maths.

Example

1   OABC is a quadrilateral. If $\overrightarrow{OA} = \mathbf{a}$, $\overrightarrow{OB} = \mathbf{b}$ and $\overrightarrow{OC} = \mathbf{b} - \dfrac{3}{4}\mathbf{a}$

Prove that $\overrightarrow{OA}$ and $\overrightarrow{CB}$ are parallel.

Answer

1   First draw a quick sketch. Don't worry about where the points A, B and C are on the diagram, just put them in any position.

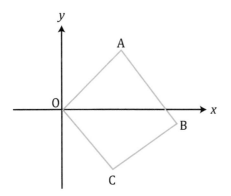

$$\overrightarrow{OA} = \mathbf{a}, \overrightarrow{OB} = \mathbf{b} \text{ and } \overrightarrow{OC} = \mathbf{b} - \frac{3}{4}\mathbf{a}$$

$$\overrightarrow{AB} = \overrightarrow{AO} + \overrightarrow{OB} = \mathbf{b} - \mathbf{a}$$

$$\overrightarrow{CB} = \overrightarrow{CO} + \overrightarrow{OA} + \overrightarrow{AB} = -\mathbf{b} + \frac{3}{4}\mathbf{a} + \mathbf{a} + \mathbf{b} - \mathbf{a}$$

$$= \frac{3}{4}\mathbf{a}$$

$\overrightarrow{CB}$ and $\overrightarrow{OA}$ both have the same vector part (i.e. $\mathbf{a}$) so they are parallel.

> Always look for vectors that are parallel (i.e. they have the same vector part). Identical vectors mean the lines are the same length and parallel to each other.

Finding angles in triangles using vectors

If you are given position vectors or vectors of lines in a triangle, they can be used to find the lengths of the sides of a triangle by working out the magnitude of each vector that is a side of the triangle. Once the lengths are known, an angle or angles can be found using the cosine rule. This technique is shown in the following step by step.

> Angles and sides in triangles involving vectors can be found using trigonometry.

Step by STEP

ABC is a triangle and the position vectors of points A and B are $5\mathbf{i} + 3\mathbf{j}$ and $12\mathbf{i} + \mathbf{j}$ respectively. If $\overrightarrow{AC} = 5\mathbf{i} + 4\mathbf{j}$, find:

1   (a)   The position vector of C.

     (b)   The size of angle ACB giving your answer to the nearest degree.

Steps to take

1   Need to find $\overrightarrow{OC}$ using a path where the vectors are known.

2   To find the angle ACB it is necessary to find the lengths of each side. To find the length of each side it will be necessary to find $\overrightarrow{BC}$ and $\overrightarrow{AB}$.

3   Find the magnitude of the vector for each side. This will give the lengths of all three sides.

4   Use the cosine rule to find the required angle.

. . . .

Answer

1 **(a)** Draw a sketch graph showing the approximate positions of points A and B.

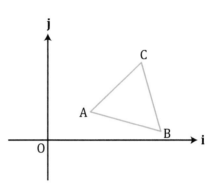

$$\vec{OC} = \vec{OA} + \vec{AC} = 5\mathbf{i} + 3\mathbf{j} + 5\mathbf{i} + 4\mathbf{j} = 10\mathbf{i} + 7\mathbf{j}$$

> |**AB**| means the magnitude of vector $\vec{AB}$ which is also the length of the side AB. The lengths of the three sides of the triangle are found in this way.

(b)　$\vec{AB} = \vec{AO} + \vec{OB} = -5\mathbf{i} - 3\mathbf{j} + 12\mathbf{i} + \mathbf{j} = 7\mathbf{i} - 2\mathbf{j}$

$\vec{BC} = \vec{BO} + \vec{OC} = -12\mathbf{i} - \mathbf{j} + 10\mathbf{i} + 7\mathbf{j} = -2\mathbf{i} + 6\mathbf{j}$

$|\vec{AB}| = \sqrt{7^2 + 2^2} = \sqrt{53}$

$|\vec{BC}| = \sqrt{(-2)^2 + 6^2} = \sqrt{40}$

$|\vec{AC}| = \sqrt{5^2 + 4^2} = \sqrt{41}$

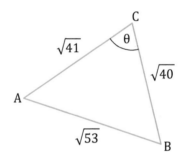

> Draw the triangle and mark on the lengths of the sides and the angle you are asked to find.

Using the cosine rule:　$\left(\sqrt{53}\right)^2 = \left(\sqrt{41}\right)^2 + \left(\sqrt{40}\right)^2 - 2\sqrt{41}\sqrt{40}\cos\theta$

$$53 = 41 + 40 - 80.99\cos\theta$$

$$\theta = 70°$$

Angle ACB = 70° (nearest degree)

Step by ⒮⒯⒠⒫

P, Q, R and S are the corners of a parallelogram.

If $\vec{OP} = 3\mathbf{j}$, $\vec{OQ} = 4\mathbf{i} + \mathbf{j}$, $\vec{OR} = 10\mathbf{i} + 5\mathbf{j}$

Using vectors, find the position vector of point S.

Steps to take

1 Draw a sketch graph with the axes labelled **i** and **j** and use the position vectors of P, Q and R to mark the approximate positions of points P, Q and R.

2 Find using the position vectors, the vectors $\vec{PQ}$ and $\vec{QR}$ in terms of **i** and **j**.

3   Use the fact that opposite sides of a parallelogram have the same vectors as they are parallel and the same length.

4   Using these, find the vector $\overrightarrow{OS}$ which is the position vector of S.

. . . .

Answer

First draw a rough sketch marking the positions of P, Q, R and S bearing in mind you are told in the question the shape is a parallelogram.

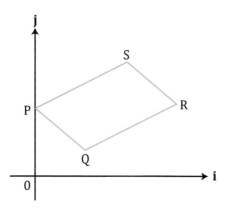

> Remember that a parallelogram has two pairs of equal parallel sides. Parallel sides of the same length will have equal vectors.

Position vector of S is $\overrightarrow{OS}$.

$$\overrightarrow{PQ} = \overrightarrow{PO} + \overrightarrow{OQ} = -3j + 4i + j = 4i - 2j$$
$$\overrightarrow{QR} = \overrightarrow{QO} + \overrightarrow{OR} = -4i - j + 10i + 5j = 6i + 4j$$
$$\overrightarrow{OS} = \overrightarrow{OQ} + \overrightarrow{QR} + \overrightarrow{RS}$$

Now $\overrightarrow{RS} = \overrightarrow{QP}$ as both vectors are parallel, the same length and pointing in the same direction. Also $\overrightarrow{QP} = -\overrightarrow{PQ}$

Hence, $\overrightarrow{OS} = \overrightarrow{OQ} + \overrightarrow{QR} - \overrightarrow{PQ} = 4i + j + 6i + 4j - 4i + 2j = 6i + 7j$

Position vector of S is 6i + 7j

Example

1   (a)   The vectors **u** and **v** are defined by **u** = 2i − 3j, **v** = −4i + 5j.

   (i)   Find the vector 4**u** − 3**v**.

   (ii)   The vectors **u** and **v** are the position vectors of the points U and V, respectively. Find the length of the line UV.   [4]

   (b)   Two villages A and B are 40 km apart on a long straight road passing through a desert. The position vectors of A and B are denoted by **a** and **b**, respectively.

   (i)   Village C lies on the road between A and B at a distance 4 km from B. Find the position vector of C in terms of **a** and **b**.

   (ii)   Village D has position vector $\frac{2}{9}\mathbf{a} + \frac{5}{9}\mathbf{b}$. Explain why village D cannot possibly be on the straight road passing through A and B.   [3]

Answer

(a) (i) $4\mathbf{u} - 3\mathbf{v} = 4(2\mathbf{i} - 3\mathbf{j}) - 3(-4\mathbf{i} + 5\mathbf{j}) = 20\mathbf{i} - 27\mathbf{j}$

(ii)

> It is always helpful to try and draw the diagram so that it approximately puts each vector in its correct position. So $-4\mathbf{i} + 5\mathbf{j}$ should be drawn four units to the left and five units up.

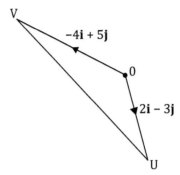

$\overrightarrow{UV} = -2\mathbf{i} + 3\mathbf{j} - 4\mathbf{i} + 5\mathbf{j} = -6\mathbf{i} + 8\mathbf{j}$

Length of line UV $= \sqrt{(-6)^2 + 8^2} = 10$

(b) (i)

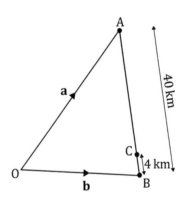

$\overrightarrow{BA} = \mathbf{a} - \mathbf{b}$ so $\overrightarrow{BC} = \dfrac{1}{10}\overrightarrow{BA} = \dfrac{1}{10}(\mathbf{a} - \mathbf{b})$

Position vector of C, $\overrightarrow{OC} = \overrightarrow{OB} + \overrightarrow{BC} = \mathbf{b} + \dfrac{1}{10}(\mathbf{a} - \mathbf{b}) = \dfrac{1}{10}\mathbf{a} + \dfrac{9}{10}\mathbf{b}$

(ii) The position vector of any point on the road is of the form

$\lambda\mathbf{a} + (1- \lambda)\mathbf{b}$ for some value of λ.

The position vector of D is $\dfrac{2}{9}\mathbf{a} + \dfrac{5}{9}\mathbf{b}$ so here $\lambda = \dfrac{2}{9}$ so $(1 - \lambda) = \dfrac{7}{9}$ so the point D does not lie on the road.

Test yourself

1 (a) Two vectors **u** and **v** are defined by **u** = 3**i** + 4**j**, **v** = −2**i** + 3**j**.
Find the vector 3**u** − 2**v**.

(b) The vectors **u** and **v** are the position vectors on the points U and V respectively. Find the exact length of the line UV.

2 Point A has position vector 4**i** − 3**j** and point B has position vector 6**i** + **j**.
(a) Find vector $\overrightarrow{AB}$.
(b) Find $|\overrightarrow{AB}|$ giving your answer as a simplified surd.

3 Shape ABCD is a square. The position vector of A, B and C are 3**i** − 2**j**, 6**i** − 4**j** and 8**i** − **j** respectively.
Find the position vector of point D.

4 Points A, B and C have coordinates (−2, 5), (4, 3) and (10, 1) respectively.
(a) Express $\overrightarrow{AB}$ and $\overrightarrow{AC}$ in terms of the unit vectors **i** and **j**.
(b) State and explain what can be deduced from your answers to part (a).
(c) State the ratio AB : AC.

5 In the diagram below, the points O, A, B, C and D are such that A is the mid-point of OD and C is the mid-point of OB.

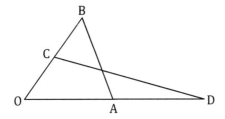

Taking O as the origin, the position vectors of A and B are denoted by **a** and **b** respectively.
(a) Show that $\overrightarrow{CD} = 2\mathbf{a} - \frac{1}{2}\mathbf{b}$.
(b) E lies on AB and is such that AE : EB = 1 : 2.
Find the position vector of E.

Summary

Condition for two vectors to be parallel

For two vectors **a** and **b** to be parallel

$\quad$ **a** = k**b** where k is a scalar

The magnitude of a vector

The vector **r** = a**i** + b**j** has magnitude given by $|\mathbf{r}| = \sqrt{a^2 + b^2}$

The position vector of a point dividing a line in a given ratio

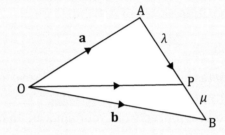

Point P dividing AB in the ratio $\lambda : \mu$ has position vector $\overrightarrow{OP}$

where $\overrightarrow{OP} = \dfrac{\mu\mathbf{a} + \lambda\mathbf{b}}{\lambda + \mu}$

Formulae

Formula booklet

The following formulae are given in the formula booklet for the AS Pure Mathematics examination.

Mensuration

Surface area of sphere = $4\pi r^2$

Area of curved surface of cone = $\pi r \times$ slant height

Binomial series

$$(a + b)^n = a^n + \binom{n}{1}a^{n-1}b + \binom{n}{2}a^{n-2}b^2 + ... + \binom{n}{r}a^{n-r}b^r + ... + b^n \qquad (n \in \mathbb{N})$$

where $\binom{n}{r} = {}^nC_r = \dfrac{n!}{r!(n-r)!}$

$$(1 + x)^n = 1 + nx + \frac{n(n-1)}{1 \times 2}x^2 + ... + \frac{n(n-1)\ ...\ (n-r+1)}{1 \times 2 \times\ ...\ r}x^r + ... \qquad \left(|x| < 1, n \in \mathbb{R}\right)$$

Vectors

The point dividing AB in the ratio $\lambda : \mu$ is $= \dfrac{\mu\mathbf{a} + \lambda\mathbf{b}}{\lambda + \mu}$

Formulae to be remembered

The following formulae are not given so need to be remembered

Mensuration

Area of a circle = πr^2

Pythagoras' theorem, $c^2 = a^2 + b^2$

Area of a triangle $= \dfrac{1}{2} \times b \times h$

Algebra

Formula for the solutions of a quadratic equation: $\quad x = \dfrac{-b \pm \sqrt{b^2 - 4ac}}{2a}$

Discriminant $= b^2 - 4ac$

Coordinate geometry

The length of a straight line joining the two points (x_1, y_1) and (x_2, y_2) is given by:
$$l = \sqrt{(x_2 - x_1)^2 + (y_2 - y_1)^2}$$

Mid-point of a line joining two points (x_1, y_1) and (x_2, y_2) is given by: $\left(\dfrac{x_1 + x_2}{2}, \dfrac{y_1 + y_2}{2}\right)$

Product of the gradients of two perpendicular lines: $m_1 m_1 = -1$

Equation of a straight line: $y - y_1 = m(x - x_1)$

Formulae

Trigonometry

$$\sin\theta = \frac{\text{opposite}}{\text{hypotenuse}}, \qquad \cos\theta = \frac{\text{adjacent}}{\text{hypotenuse}}, \qquad \tan\theta = \frac{\text{opposite}}{\text{adjacent}}$$

The Sine rule: $\quad \dfrac{a}{\sin A} = \dfrac{b}{\sin B} = \dfrac{c}{\sin C}$

The Cosine rule: $\quad a^2 = b^2 + c^2 - 2bc\cos A$

Area of triangle $= \dfrac{1}{2}ab\sin C$

$\tan\theta = \dfrac{\sin\theta}{\cos\theta}$ and $\cos^2\theta + \sin^2\theta = 1$

Differentiation

If $y = kx^n$, $\dfrac{dy}{dx} = nkx^{n-1}$

Integration

$\int x^n dx = \dfrac{x^{n+1}}{n+1} + c \quad$ (provided $n \neq -1$)

Vectors

If $\mathbf{r} = a\mathbf{i} + b\mathbf{j}, \quad |\mathbf{r}| = \sqrt{a^2 + b^2}$

Test yourself answers

Topic 1

1 Since there is a finite number of numbers we can try each integer in turn therefore using proof by exhaustion.

n	$n^2 + 2$	Divisible by 4?
2	$2^2 + 2 = 6$	No
3	$3^2 + 2 = 11$	No
4	$4^2 + 2 = 18$	No
5	$5^2 + 2 = 27$	No
6	$6^2 + 2 = 38$	No
7	$7^2 + 2 = 51$	No

Since all the allowable values for n have been used and the resulting value is not divisible by 4, we have proved by exhaustion that $n^2 + 2$ is not divisible by 4.

2 If $a = 3$, then $|a + 1| = |3 + 1| = 4$

If $b = -5$, then $|-5 + 1| = |-4| = 4$

Here $|a + 1| = |b + 1|$ but $a \neq b$ so the statement is false.

3 Let the consecutive integers be $n, n + 1, n + 2, n + 3$

Product of the last two integers:

$$(n + 2)(n + 3) = n^2 + 5n + 6$$

Product of the first two integers:

$$n(n + 1) = n^2 + n$$

Difference in products $= (n^2 + 5n + 6) - (n^2 + n)$

$$= 4n + 6$$

Sum of the four integers:

$$n + n + 1 + n + 2 + n + 3 = 4n + 6$$

These are both equal so the statement has been proved.

4 We only need to find two values of a and b for which the inequality does not hold. As we need to find $\sqrt{ab}$ it makes sense to use values for a and b that will make the product a square number.

Try $a = -20$ and $b = -5$.

$a + b = -20 + (-5) = -25$ and $2\sqrt{ab} = 2\sqrt{(-20) \times (-5)}$

$$= 2\sqrt{100} = \pm 20$$

Now in this case $-25 < 20$ and $-25 < -20$

So a counter-example has been found where the proposition is not true so the proposition is false.

5 All numbers can be expressed in the following ways using multiples of 3:

$p = 3n$ would give 0, 3, 6, 9, 12, ...

$p = 3n + 1$ would give 1, 4, 7, 10, 13, ...

$p = 3n + 2$ would give 2, 5, 8, 11, 14, ...

If $p = 3n$, $p^2 = 9n^2$ which is a multiple of 3.

If $p = 3n + 1$, $p^2 = (3n + 1)^2 = 9n^2 + 6n + 1 = 3(3n^2 + 2n) + 1$ which is one more than a multiple of 3.
If $p = 3n + 2$, $p^2 = (3n + 2)^2 = 9n^2 + 12n + 4 = 3(3n^2 + 4n + 1) + 1$ which is one more than a multiple of 3.
Hence the statement has been proved true.

6 (a) If $a = -1$ and $b = -2$, then $a^2 = 1$ and $b^2 = 4$.
For this example $b^2 > a^2$ yet $b < a$.
Hence, by this counter-example, the statement is disproved.

(b) First try some ordinary positive numbers for x, y and z such that $x > y$.
Let $x = 2$, $y = 1$ and $z = 3$.
Hence, $xz = 6$ and $yz = 3$
So in this case $xz > yz$ (so as this is true; it is not a counter-example).
Now try combinations of positive and negative numbers such that $x > y$.
Let $x = 8$, $y = 6$ and $z = -3$.
Hence, $xz = -24$ and $yz = -18$
$-24 < -18$ so $xz < yz$, so the counter-example has disproved the statement.

7 If $n = 6$, then $n^2 = 36$ and n^2 is divisible by 4 exactly.
However, 6 is not divisible by 4 exactly.
Hence the statement is incorrect due to this counter-example.

Topic 2

1 $y = 5x^{\frac{1}{2}} + 45x^{-1} - 7$

2 (a) 1

(b) $\frac{1}{9}$

(c) 2

(d) $\frac{1}{5}$

(c) 64

3 (a) $\sqrt{48} + \frac{12}{\sqrt{3}} - \sqrt{27} = 4\sqrt{3} + \frac{12\sqrt{3}}{\sqrt{3}\sqrt{3}} - 3\sqrt{3}$

$= 4\sqrt{3} + 4\sqrt{3} - 3\sqrt{3} = 5\sqrt{3}$

(b) $\frac{2 + \sqrt{5}}{3 + \sqrt{5}} = \frac{(2 + \sqrt{5})(3 - \sqrt{5})}{(3 + \sqrt{5})(3 - \sqrt{5})} = \frac{6 + \sqrt{5} - 5}{9 - 5}$

$= \frac{1 + \sqrt{5}}{4}$

4 As this question is about the nature of roots, we first find the discriminant
$b^2 - 4ac = (5)^2 - 4(k)(-7) = 25 + 28k$
For no real roots, $b^2 - 4ac < 0$
Hence $25 + 28k < 0$
So, $28k < -25$, giving $k < -\frac{25}{28}$

5 $x^2 - 6x + 8 > 0$
$(x - 4)(x - 2) > 0$

As the curve $y = x^2 - 6x + 8$ has a positive coefficient of x^2 the curve will be U-shaped, cutting the x-axis at $x = 4$ and $x = 2$.
Sketching the curve for $y = x^2 - 6x + 8$ gives the following:

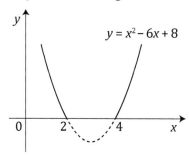

We want the part of the graph which is above the x-axis.
The range of values for which this occurs is $x < 2$ or $x > 4$

6 $5x^2 - 20x + 10 = 5(x^2 - 4x + 2)$
$$= 5\left[(x - 2)^2 - 4 + 2\right]$$
$$= 5(x - 2)^2 - 10$$
Giving $a = 5$, $b = -2$, $c = -10$

7 $1 - 3x < x + 7$
$-3x < x + 6$

The inequality sign is reversed because both sides have been divided by a negative quantity (i.e. -4).

$-4x < 6$
$$x > -\frac{6}{4}$$
$$x > -\frac{3}{2}$$

BOOST
Grade ⇧⇧⇧⇧

If you do not cancel fractions you may lose marks.

8 $y = x + 4$ and $y = x^2 - 7x + 20$
So $x + 4 = x^2 - 7x + 20$
$x^2 - 8x + 16 = 0$
$(x - 4)(x - 4) = 0$
$(x - 4)^2 = 0$

There is only one solution to the quadratic which means the straight line and curve touch at only one point.

If the line and the curve touch then the resulting equation will have a repeated root.

There is one repeated solution to this equation which proves that the straight line and curve touch.
Solving gives $x = 4$
Substituting $x = 4$ into the equation of the straight line,
$y = 4 + 4 = 8$
Hence, the coordinates of the point of contact are $(4, 8)$
An alternative method for proving that the curve and straight line touch at one point is to find the discriminant and show that it equals zero.
For example the equation $x^2 - 8x + 16 = 0$ has discriminant
$b^2 - 4ac = (-8)^2 - 4(1)(16) = 64 - 64 = 0$.
This shows there are two real equal roots showing the curve and straight line touch at a single point.

9 Let $f(x) = 4x^3 + 3x^2 - 3x + 1$
$f(-1) = 4(-1)^3 + 3(-1)^2 - 3(-1) + 1 = 3$
Therefore remainder $= 3$

10 (a) Let $f(x) = x^3 + 6x^2 + ax + 6$
$f(-2) = (-2)^3 + 6(-2)^2 + a(-2) + 6 = 22 - 2a$
If $x + 2$ is a factor, $f(-2) = 0$
Hence, $22 - 2a = 0$
So $a = 11$

(b) $(x + 2)(ax^2 + bx + c) = x^3 + 6x^2 + 11x + 6$
Equating coefficients of x^3 gives $a = 1$.
Equating coefficients of x^2 gives $b + 2a = 6$ and since $a = 1$ this gives $b = 4$.
Equating constant terms gives $2c = 6$, so $c = 3$.
So $(x + 2)(x + 1)(x + 3) = 0$
Solving gives x = –2, –1 or –3

11 Need to first find the equation of the line passing through Q (2, 0) and R (0, 1).
Gradient of line $= -\frac{1}{2}$ and intercept on y-axis, $c = 1$.
Equation of the line is $y = -\frac{1}{2}x + 1$
Inequalities are: $b < -\frac{1}{2}a + 1$, $b > a^2 - 4$, $a > 0$

12 This will be a cubic graph and since the coefficient (i.e. the number in front of) of x^3 is positive, the graph will look like this.

$0 = (x + 4)(x - 2)(x - 7)$ so the curve will intersect the x-axis at –4, 2 and 7.
When $x = 0$ (i.e the equation of the y-axis), $y = (0 + 4)(0 - 2)(0 - 7)$ so the curve intersects the y-axis at $y = 56$.
The sketch can now be drawn

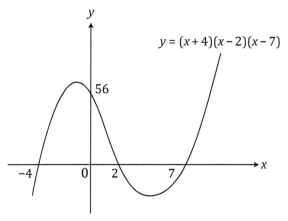

This is obtained by completing the square.

13 $x^2 - 1.2x - 3.64 = (x - 0.6)^2 - 0.36 - 3.64$
$= (x - 0.6)^2 - 4$
Hence $p = 0.6$
$x^2 - 1.2x - 3.64 = 0$
So $(x - 0.6)^2 - 4 = 0$
$(x - 0.6)^2 = 4$
$x - 0.6 = \pm 2$
$x = 2 + 0.6$ or $x = -2 + 0.6$
Hence x = 2.6 or x = –1.4

14 For distinct and real roots, $b^2 - 4ac > 0$
Hence $k^2 - 4(k - 1)(k) > 0$
$k^2 - 4k^2 + 4k > 0$
$-3k^2 + 4k > 0$
$3k^2 - 4k < 0$

Remember to reverse the inequality when dividing by –1.

Factorising gives $k(3k - 4) < 0$

If a graph of $y = k(3k - 4)$ were plotted with values of k on the x-axis, as there is a positive coefficient of k^2 the curve will be U-shaped, cutting the x-axis at $k = 0$ and $k = \frac{4}{3}$.

Without drawing the graph you can see that the section of the graph needed will be below the x-axis.

Hence the required range of k is $0 < k < \frac{4}{3}$

15 $4x^2 - 12x + 9 = 4\left[x^2 - 3x + \frac{9}{4}\right]$

$$= 4\left[\left(x - \frac{3}{2}\right)^2 - \frac{9}{4} + \frac{9}{4}\right]$$

$$= 4\left(x - \frac{3}{2}\right)^2$$

Comparing the above expression with $a(x + b)^2 + c$, gives:

$a = 4,$ $b = -\frac{3}{2},$ $c = 0$

The graph of $4\left(x - \frac{3}{2}\right)^2$ is U-shaped because the coefficient of x^2 is positive.

$4\left(x - \frac{3}{2}\right)^2$ has its minimum point when $x = \frac{3}{2}$. When $x = \frac{3}{2}, y = 0$.

Hence the coordinates of the stationary point are $\left(\frac{3}{2}, 0\right)$.

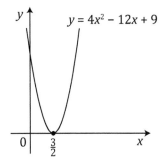

16 Factorising $x^2 - 2x - 15$ gives

$(x - 5)(x + 3) = 0$

Hence $x = 5$ or $x = -3$

As the coefficient of x^2 is positive the graph of $x^2 - 2x - 15$ is U-shaped.

Now $x^2 - 2x - 15 \leq 0$. This is the region below the x-axis (i.e. where $y \leq 0$).

Hence $-3 \leq x \leq 5$

17 (a) $y = -f(x)$

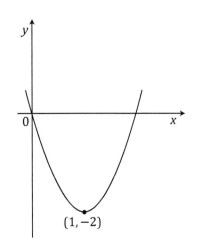

This transformation represents a reflection in the x-axis.

This transformation represents a one-way stretch of scale factor 3, parallel to the y-axis.

(b) $y = 3f(x)$

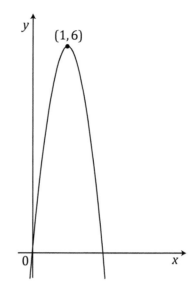

This transformation represents a translation of $\begin{pmatrix} 1 \\ 0 \end{pmatrix}$

(c) $y = f(x - 1)$

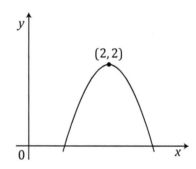

This transformation represents a one-way stretch with scale factor $\frac{1}{2}$ parallel to the x-axis

(d) $y = f(2x)$

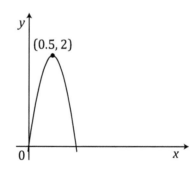

Topic 3

① (a) Gradient of AB $= \dfrac{y_2 - y_1}{x_2 - x_1} = \dfrac{1 - 0}{4 - 1} = \dfrac{1}{3}$

Gradient of CD $= \dfrac{y_2 - y_1}{x_2 - x_1} = \dfrac{4 - 3}{2 - (-1)} = \dfrac{1}{3}$

As the gradients of AB and CD are the same the two lines are parallel.

(b) Gradient of AB $= \frac{1}{3}$ and AB passes through point A $(1, 0)$ so equation of AB is:

$$y - y_1 = m(x - x_1)$$

$$y - 0 = \frac{1}{3}(x - 1)$$
$$3y = x - 1$$

Rearranging this equation so that it is in the form asked for by the question gives:
$$x - 3y - 1 = 0$$

2 (a) Gradient of AB $= \dfrac{y_2 - y_1}{x_2 - x_1} = \dfrac{-1 - 4}{k - (-7)} = \dfrac{-5}{k + 7}$

But gradient of AB $= -\dfrac{1}{2}$ so $\dfrac{-5}{k + 7} = -\dfrac{1}{2}$

$$-5 \times 2 = -1(k + 7)$$
$$-10 = -k - 7$$

Giving $\qquad k = 3$

> The equation is multiplied through by the common denominator, $2(k + 7)$.

(b) The product of the gradients of perpendicular lines is −1. Hence,

$$m\left(-\tfrac{1}{2}\right) = -1$$

Hence gradient of BC = 2

Equation BC is:
$$y - y_1 = m(x - x_1) \text{ where } m = 2 \text{ and } (x_1, y_1) = (3, -1).$$
$$y - (-1) = 2(x - 3)$$
$$y + 1 = 2x - 6$$
$$2x - y - 7 = 0$$

3 (a) Gradient of AB $= \dfrac{y_2 - y_1}{x_2 - x_1} = \dfrac{6 - 2}{1 - (-3)} = \dfrac{4}{4} = 1$

Gradient of BC $= \dfrac{y_2 - y_1}{x_2 - x_1} = \dfrac{1 - 6}{6 - 1} = \dfrac{-5}{5} = -1$

Product of gradients = (1)(−1) = −1 proving that the two lines are perpendicular to each other.

(b) $\sqrt{(x_2 - x_1)^2 + (y_2 - y_1)^2}$

Substituting the coordinates A (−3, 2) and B (1, 6) into the formula gives:

$$AB = \sqrt{[1 - (-3)]^2 + (6 - 2)^2} = \sqrt{16 + 16} = \sqrt{32} \text{ units}$$

Using the coordinates B (1, 6) and C (6, 1) in the formula gives:

$$BC = \sqrt{(6 - 1)^2 + (1 - 6)^2} = \sqrt{25 + 25} = \sqrt{50} \text{ units}$$

(c) $\tan A\hat{C}B = \dfrac{AB}{BC} = \dfrac{\sqrt{32}}{\sqrt{50}} = \dfrac{\sqrt{16 \times 2}}{\sqrt{25 \times 2}} = \dfrac{4\sqrt{2}}{5\sqrt{2}} = \dfrac{4}{5}$

> AS Pure material is needed here to enable the surds to be simplified.

4 (a) Gradient of AB $= \dfrac{y_2 - y_1}{x_2 - x_1} = \dfrac{4 - (-1)}{-7 - 3} = -\dfrac{1}{2}$

(b) Equation of straight line which passes through (−7, 4) and has gradient $-\dfrac{1}{2}$ is given by:

$$y - y_1 = m(x - x_1) \text{ where } m = -\tfrac{1}{2} \text{ and } (x_1, y_1) = (-7, 4).$$
$$y - 4 = -\tfrac{1}{2}(x - (-7))$$
$$2y - 8 = -x - 7$$
$$x + 2y - 1 = 0$$

> **BOOST**
> **Grade** ⇧⇧⇧⇧
>
> This formula needs to be remembered. If you forget it you can plot the two points on a sketch graph, form a triangle and use Pythagoras' theorem to find the length of the hypotenuse.
>
> Watch the signs when using this formula. Add brackets to the negative coordinates to emphasise them.

(c) The length of a straight line joining the two points (x_1, y_1) and (x_2, y_2) is given by:

$$\sqrt{(x_2 - x_1)^2 + (y_2 - y_1)^2}$$

Substituting the coordinates $(-7, 4)$ and $(3, -1)$ into this gives:

Length of AB $= \sqrt{[3 - (-7)]^2 + (-1 - 4)^2} = \sqrt{100 + 25} = \sqrt{125} = 5\sqrt{5}$ units

(d) The mid-point of a line joining the points (x_1, y_1) and (x_2, y_2) is given by:

$$\left(\frac{x_1 + x_2}{2}, \frac{y_1 + y_2}{2}\right)$$

Hence mid-point E of AB $= \left(\frac{-7 + 3}{2}, \frac{4 + (-1)}{2}\right) = \left(-2, \frac{3}{2}\right)$

(e) If CD is perpendicular to AB then the product of the gradients equals -1.

$$m_1 m_2 = -1$$

$$\left(-\frac{1}{2}\right)m_2 = -1$$

Giving gradient of CD $= 2$

Finding the gradients using the coordinates of C $(6, 1)$ and D $(k, -15)$ gives:

Gradient of AB $= \dfrac{y_2 - y_1}{x_2 - x_1} = \dfrac{-15 - 1}{k - 6}$

The gradient of CD is 2 so an equation can be formed.

Hence $\qquad \dfrac{-15 - 1}{k - 6} = 2$

$$-16 = 2(k - 6)$$
$$-16 = 2k - 12$$
$$k = -2$$

5 (a) Comparing the equation $x^2 + y^2 - 8x - 6y = 0$ with the equation
$x^2 + y^2 + 2gx + 2fy + c = 0$

Alternatively, you could use the method of completing the square here.

we can see $g = -4$, $f = -3$, $c = 0$.
Centre A has coordinates $(-g, -f) = (4, 3)$
Radius $= \sqrt{g^2 + f^2 - c} = \sqrt{(-4)^2 + (-3)^2 - 0} = \sqrt{25} = 5$

(b) Rearranging the equation of the straight line for y gives
$y = -2x - 4$
Substituting y into the equation of the circle gives:
$$x^2 + (-2x - 4)^2 - 8x - 6(-2x - 4) = 0$$
$$x^2 + 4x^2 + 16x + 16 - 8x + 12x + 24 = 0$$
$$5x^2 + 20x + 40 = 0$$
$$x^2 + 4x + 8 = 0$$
Discriminant $= b^2 - 4ac = 16 - 4 \times 1 \times 8 = 16 - 32 = -16$
As $b^2 - 4ac < 0$ there are no real roots so the circle and line do not intersect.

6 (a) Comparing the equation $x^2 + y^2 - 4x + 6y - 3 = 0$ with the equation
$x^2 + y^2 + 2gx + 2fy + c = 0$, we can see $g = -2$, $f = 3$, $c = -3$.
Centre A has coordinates $(-g, -f) = (2, -3)$

Alternatively, you could use the method of completing the square here.

Radius $= \sqrt{g^2 + f^2 - c} = \sqrt{(-2)^2 + (3)^2 - (-3)} = \sqrt{16} = 4$

(b) If point P $(2, 1)$ lies on the circle its coordinates will satisfy the equation of the circle $x^2 + y^2 - 4x + 6y - 3 = 0$.
$$x^2 + y^2 - 4x + 6y - 3 = (2)^2 + (1)^2 - 4(2) + 6(1) - 3 = 4 + 1 - 8 + 6 - 3 = 0$$
Both sides of the equation equal zero so P $(2, 1)$ lies on the circle.

7 (a) Equation of the circle is:
$$(x - a)^2 + (y - b)^2 = r^2$$
$$(x - 2)^2 + (y - 3)^2 = 25$$
$$x^2 - 4x + 4 + y^2 - 6y + 9 = 25$$
$$x^2 + y^2 - 4x - 6y - 12 = 0$$

(b) Gradient of line joining the centre A (2, 3) to P (5, 7) $= \dfrac{7 - 3}{5 - 2} = \dfrac{4}{3}$

Gradient of tangent $= -\dfrac{3}{4}$

Equation of tangent is: $y - 7 = -\dfrac{3}{4}(x - 5)$

$$4y - 28 = -3x + 15$$
$$3x + 4y - 43 = 0$$

8 (a) $x^2 + y^2 - 8x + 2y + 7 = 0$
Completing the square gives:
$$(x - 4)^2 + (y + 1)^2 - 16 - 1 + 7 = 0$$
This gives $\quad (x - 4)^2 + (y + 1)^2 = 10$
Comparing this with $(x - a)^2 + (y - b)^2 = r^2$ gives centre A(4, −1) and radius $= \sqrt{10}$.

(b) (i) If point P lies on the circle, its coordinates will satisfy the equation of the circle.
Substituting the coordinates (7, −2) into the LHS of the equation:
LHS $= 7^2 + (-2)^2 - 8(7) + 2(-2) + 7 = 49 + 4 - 56 - 4 + 7 = 0 = $ RHS
The coordinates of P satisfy the equation, proving that P lies on the circle.

(ii) A(4, −1) and P(7, −2)

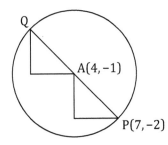

As AP = AQ to go from P to A the *x*-coordinate decreases by 3 and the *y*-coordinate increases by 1. We can use this fact when going from A to Q. Decreasing the *x*-coordinate of A by 3 gives 1 and increasing the *y*-coordinate by 1 gives 0. Hence the coordinates of Q are (1, 0).

(c) $y = 2x - 4$ is substituted into the equation of the circle.
$$x^2 + y^2 - 8x + 2y + 7 = 0$$
$$x^2 + (2x - 4)^2 - 8x + 2(2x - 4) + 7 = 0$$
$$x^2 + 4x^2 - 16x + 16 - 8x + 4x - 8 + 7 = 0$$
$$5x^2 - 20x + 15 = 0$$
$$x^2 - 4x + 3 = 0$$
$$(x - 3)(x - 1) = 0$$

$x = 3 \quad$ or $\quad x = 1$
When $x = 3, y = 2(3) - 4 = 2$
When $x = 1, y = 2(1) - 4 = -2$
Hence, points of intersection are (3, 2) and (1, −2)

An alternative method of verifying that P lies on C involves showing that the distance of P from the centre A is the same as the radius of the circle, as follows:
$$AP^2 = (x_1 - x_2)^2 + (y_1 - y_2)^2$$
$$= (7 - 4)^2 + (-2 - (-1))^2$$
$$= 3^2 + (-1)^2 = 9 + 1$$
$$= 10 = r^2 \Rightarrow AP = r$$

The distance from the centre of the circle A to the point P is equal to the radius of C, therefore P lies on C.

Divide through by 5 to simplify this quadratic equation.

Each *x*-coordinate is substituted into the equation of the straight line to find the corresponding *y*-coordinate.

9   (a)   (i)   Centre of circle is at the mid-point of the diameter PQ.

Mid-point of line joining P (1, –4) and Q (9, 10) is:

$$\left(\frac{1+9}{2}, \frac{-4+10}{2}\right) = (5, 3)$$

(ii)   Distance between the points (1, –4) and (5, 3) is given by:

> The distance from the mid-point to the circumference is the radius of the circle.

$$r = \sqrt{(x_2 - x_1)^2 + (y_2 - y_1)^2}$$
$$r = \sqrt{(5 - 1)^2 + [3 - (-4)]^2}$$
$$r = \sqrt{4^2 + 7^2}$$
$$r = \sqrt{16 + 49}$$
$$r = \sqrt{65}$$

(iii)   The equation of a circle having centre (a, b) and radius r is given by

$$(x - a)^2 + (y - b)^2 = r^2$$

For this circle, centre is (5, 3) and radius is $\sqrt{65}$.

$$(x - 5)^2 + (y - 3)^2 = 65$$

(b)   Substituting the coordinates of R into the LHS of the equation gives:

LHS = $(4 - 5)^2 + (11 - 3)^2 = (-1)^2 + (8)^2 = 65 = $ RHS

so the coordinates of R lie on the circle.

(c)   PQ is a diameter and as R is a point on the circumference, angle $P\widehat{R}Q = 90°$ because it is an angle in a semicircle.

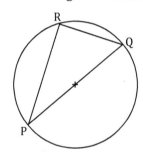

> The formula for the distance between two points is used here.

Length of PQ = $2r = 2\sqrt{65}$

Length of QR = $\sqrt{(x_2 - x_1)^2 + (y_2 - y_1)^2}$

$= \sqrt{(9 - 4)^2 + (10 - 11)^2}$

$= \sqrt{26}$

Using trigonometry $\sin Q\widehat{P}R = \dfrac{QR}{PQ} = \dfrac{\sqrt{26}}{2\sqrt{65}}$

$= 0.3162$

Hence $Q\widehat{P}R = \sin^{-1}(0.3162) = 18.4°$

Topic 4

> The value of $\binom{5}{2}$ could be found using Pascal's triangle by looking along the row which starts at 1 and then 5. The third number in the row (i.e. 10) gives the value of $\binom{5}{2}$.

1   Obtaining the formulae from the formula booklet:

$$(a + b)^n = a^n + \binom{n}{1}a^{n-1}b + \binom{n}{2}a^{n-2}b^2 + ...$$

$$\binom{n}{r} = {}^nC_r = \frac{n!}{r!(n - r)!}$$

The term x^2 is given by $\binom{n}{2}a^{n-2}b^2$

Here $a = 2$, $b = 3x$ and $n = 5$

So the term in x^2 is $\dfrac{5!}{2!(5-2)!}(2)^3(3x)^2 = 10 \times 8 \times 9x^2 = 720x^2$

Hence, the coefficient of x^2 is 720

2 From the formula booklet:

$(1 + x)^n = 1 + nx + \dfrac{n(n-1)}{2!}x^2 + \dfrac{n(n-1)(n-2)}{3!}x^3 + \ldots$

In this case we substitute $3x$ for x and 6 for n.

Hence $(1 + 3x)^6 \approx 1 + (6)(3x) + \dfrac{(6)(5)}{2 \times 1}(3x)^2 + \dfrac{(6)(5)(4)}{3 \times 2 \times 1}(3x)^3$

$\approx 1 + 18x + 135x^2 + 540x^3$

3 Obtaining the formula gives:

$(a + b)^n = a^n + \binom{n}{1}a^{n-1}b + \binom{n}{2}a^{n-2}b^2 + \ldots$

Here $n = 3$, $a = 3$, $b = 2x$

$(3 + 2x)^3 = 3^3 + 3(3)^2(2x) + \dfrac{(3)(2)}{2!}3^1(2x)^2 + \dfrac{(3)(2)(1)}{3!}3^0(2x)^3$

$= 27 + 54x + 36x^2 + 8x^3$

4 Using the binomial expansion for $(a + b)^5$ we obtain:

$(a + b)^5 = a^5 + 5a^4b + 10a^3b^2 + 10a^2b^3 + 5ab^4 + b^5$

Substituting $a = \sqrt{5}$ and $b = -\sqrt{2}$ into the above, gives:

$(\sqrt{5} - \sqrt{2})^5 = (\sqrt{5})^5 + 5(\sqrt{5})^4(-\sqrt{2}) + 10(\sqrt{5})^3(-\sqrt{2})^2 + 10(\sqrt{5})^2(-\sqrt{2})^3 + 5(\sqrt{5})(-\sqrt{2})^4 + (-\sqrt{2})^5$

$= 25\sqrt{5} - 125\sqrt{2} + 100\sqrt{5} - 100\sqrt{2} + 20\sqrt{5} - 4\sqrt{2}$

$= 145\sqrt{5} - 229\sqrt{2}$

Hence, $a = 145$ and $b = -229$

5 (a) $(1 + x)^n = 1 + nx + \dfrac{n(n-1)x^2}{2!} + \dfrac{n(n-1)(n-2)x^3}{3!} + \ldots$

$\left(1 + \sqrt{6}\right)^5 = 1 + 5\sqrt{6} + \dfrac{5(4)\left(\sqrt{6}\right)^2}{2!} + \dfrac{5(4)(3)\left(\sqrt{6}\right)^3}{3!} + \dfrac{5(4)(3)(2)\left(\sqrt{6}\right)^4}{4!}$

$+ \dfrac{5(4)(3)(2)(1)\left(\sqrt{6}\right)^5}{5!}$

$= 1 + 5\sqrt{6} + 60 + 60\sqrt{6} + 180 + 36\sqrt{6}$

$= 241 + 101\sqrt{6}$

Giving $a = 241$ and $b = 101$

(b) $(1 + x)^n = 1 + nx + \dfrac{n(n-1)x^2}{2!} + \dfrac{n(n-1)(n-2)x^3}{3!} + \ldots$

Term in x^2 is $\dfrac{n(n-1)\,x^2}{2!}$ but here x is substituted by $3x$

Term in x^2 is $\dfrac{n(n-1)(3x)^2}{2!} = \dfrac{n(n-1)9x^2}{2}$

But the coefficient of x^2 is 495, so

$\dfrac{n(n-1)9}{2} = 495$

$n^2 - n - 110 = 0$

$(n - 11)(n + 10) = 0$

n is positive, so $n = 11$

Topic 5

1 (a) Area $= \frac{1}{2}bc \sin A = \frac{1}{2} \times 12 \times 8 \times \sin 150° = 24\,\text{cm}^2$

(b) Using cosine rule
$$a^2 = b^2 + c^2 - 2bc \cos A$$
$$= 12^2 + 8^2 - 2 \times 12 \times 8 \cos 150° = 374.2769$$
$$a = 19.3462$$
$$a = 19.3\,\text{cm (to one decimal place)}$$

> Remember not to round off answers to the required number of decimal places until the final answer.

2 (a) $(0, 0), (\pi, 0), (2\pi, 0), (3\pi, 0), (4\pi, 0)$

(b) $\left(\frac{\pi}{2}, 1\right), \left(\frac{3\pi}{2}, -1\right), \left(\frac{5\pi}{2}, 1\right), \left(\frac{7\pi}{2}, -1\right)$

3

> No diagram is given in the question, so draw your own and add the information. You can now see the pairing of the sides and angles which points to the need to use the sine rule.

Using the sine rule
$$\frac{\sin C}{c} = \frac{\sin A}{a}$$
$$\frac{\sin A\widehat{C}B}{4} = \frac{\sin 30°}{3\sqrt{2} - 1}$$
$$\sin A\widehat{C}B = \frac{4 \sin 30°}{3\sqrt{2} - 1}$$

> Remove the surd from the denominator by multiplying the numerator and denominator by the conjugate of the denominator, i.e. $(3\sqrt{2} + 1)$.

$$\sin A\widehat{C}B = \frac{2}{3\sqrt{2} - 1} = \frac{2(3\sqrt{2} + 1)}{(3\sqrt{2} - 1)(3\sqrt{2} + 1)}$$
$$\sin A\widehat{C}B = \frac{6\sqrt{2} + 2}{18 - 1} = \frac{2 + 6\sqrt{2}}{17}$$

4 $3 \sin^2 \theta = 5 - 5 \cos \theta$
$$3(1 - \cos^2 \theta) = 5 - 5 \cos \theta$$
$$3 - 3 \cos^2 \theta = 5 - 5 \cos \theta$$
$$3 \cos^2 \theta - 5 \cos \theta + 2 = 0$$
$$(\cos \theta - 1)(3 \cos \theta - 2) = 0$$

> Use the trig identity $\cos^2\theta + \sin^2\theta = 1$ to create a quadratic equation in $\cos\theta$ which can then be solved.

$\cos \theta = 1 \quad$ or $\quad \cos \theta = \frac{2}{3}$

$\theta = \cos^{-1}(1) = 0°, 360°, \quad$ or $\quad \theta = \cos^{-1}\left(\frac{2}{3}\right) = 48.2°, (360 - 48.2)°$

Hence $\theta = 0°, 48.2°, 311.8°$ or $360°$

5 (a) Using cosine rule, we have:
$$(x + 2)^2 = x^2 + (x - 2)^2 - 2 \times x \times (x - 2) \cos B\widehat{A}C$$
$$x^2 + 4x + 4 = x^2 + x^2 - 4x + 4 - 2x(x - 2) \cos B\widehat{A}C$$
$$-x^2 + 8x = -2x(x - 2) \cos B\widehat{A}C$$
$$\cos B\widehat{A}C = \frac{x(8 - x)}{-2x(x - 2)} = \frac{-(8 - x)}{2(x - 2)} = \frac{x - 8}{2x - 4}$$

(b) $\cos 120° = -\dfrac{1}{2}$, hence $\dfrac{x-8}{2x-4} = -\dfrac{1}{2}$

$2x - 16 = -2x + 4$
Hence, $x = 5$ cm

Substitute $x = 5$ to find the actual lengths of the sides.

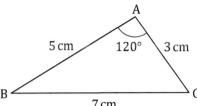

Using the sine rule to find $A\widehat{B}C$:

$\dfrac{\sin A\widehat{B}C}{3} = \dfrac{\sin 120°}{7}$

$A\widehat{B}C = 21.8°$

6 (a) $4\cos^2\theta + 1 = 4\sin^2\theta - 2\cos\theta$

$4\cos^2\theta + 1 = 4(1 - \cos^2\theta) - 2\cos\theta$

$4\cos^2\theta + 1 = 4 - 4\cos^2\theta - 2\cos\theta$

$8\cos^2\theta + 2\cos\theta - 3 = 0$

$(4\cos\theta + 3)(2\cos\theta - 1) = 0$

$\cos\theta = -\dfrac{3}{4}$ or $\cos\theta = \dfrac{1}{2}$

When $\cos\theta = -\dfrac{3}{4}$, $\theta = 138.59°$ or $221.41°$

When $\cos\theta = \dfrac{1}{2}$, $\theta = 60°$ or $300°$

(b) $\sin(\alpha + 40°) = \dfrac{1}{\sqrt{2}}$

So as $\sin 45°$ *or* $\sin 135° = \dfrac{1}{\sqrt{2}}$,

so $\alpha + 40° = 45°$ *or* $135°$

Hence $\alpha = 5°$ *or* $95°$

$\sin(\alpha - 35°) = \dfrac{\sqrt{3}}{2}$

So as $\sin 60°$ *or* $\sin 120° = \dfrac{\sqrt{3}}{2}$,

so $\alpha - 35° = 60°$ *or* $120°$

Hence $\alpha = 95°$ *or* $155°$

As α has to satisfy *both* of the equations, $\alpha = 95°$

(c) $\dfrac{7}{\cos\phi} - \dfrac{10}{\sin\phi} = 0$

so $\dfrac{7}{\cos\phi} = \dfrac{10}{\sin\phi}$

$\dfrac{\sin\phi}{\cos\phi} = \dfrac{10}{7}$

so $\tan\phi = \dfrac{10}{7}$

$\phi = 55°, 235°$

Topic 6

1 $\log_2 36 - 2\log_2 15 + \log_2 100$
$$= \log_2 36 - \log_2 15^2 + \log_2 100$$
$$= \log_2 36 - \log_2 225 + \log_2 100$$
$$= \log_2 \left(\frac{36 \times 100}{225}\right)$$
$$= \log_2 16$$

The squaring or cube rooting can be done in either order. It is easier, however, to cube root first and then square.

2 $\log_{27} x = \frac{2}{3}$
$$x = 27^{\frac{2}{3}} = \sqrt[3]{27^2} = 3^2$$
So $x = 9$

3 $3^x = 2$
Taking logs to base 10 of both sides
$$\log 3^x = \log 2$$
$$x \log 3 = \log 2$$
$$x = \frac{\log 2}{\log 3}$$
$$x = 0.63 \ (2 \text{ d.p.})$$

4 $\frac{1}{2}\log_a 36 - 2\log_a 6 + \log_a 4$
$$= \log_a 36^{\frac{1}{2}} - \log_a 6^2 + \log_a 4$$
$$= \log_a 6 - \log_a 36 + \log_a 4$$
$$= \log_a \left(\frac{6 \times 4}{36}\right)$$
$$= \log_a \left(\frac{2}{3}\right)$$

5 $\log_a (6x^2 + 5) - \log_a x = \log_a 17$
$$\log_a \left(\frac{6x^2 + 5}{x}\right) = \log_a 17$$
$$\frac{6x^2 + 5}{x} = 17$$
$$6x^2 + 5 = 17x$$
$$6x^2 - 17x + 5 = 0$$
$$(3x - 1)(2x - 5) = 0$$
Giving $x = \frac{1}{3}$ or $x = \frac{5}{2}$

6 Let $p = \log_a 19$ and $q = \log_7 a$
Hence $a^p = 19$ and $7^q = a$
So $(7^q)^p = 19$
Hence $7^{pq} = 19$
So $pq = \log_7 19$
As $p = \log_a 19$ and $q = \log_7 a$, $pq = \log_a 19 \times \log_7 a$
Hence $\log_7 a \times \log_a 19 = \log_7 19$

7 $6^x = 12$

Taking logs of both sides:

$$\log 6^x = \log 12$$
$$x \log 6 = \log 12$$
$$x = \frac{\log 12}{\log 6}$$
$$x = 1.387 \ (3 \text{ d.p.})$$

> The rule $\log_a x^k = k \log_a x$ is used here.

8 $9^x - 6 \times 3^x + 8 = 0$

Now $9^x = (3^2)^x = 3^{2x} = (3^x)^2$

$$(3^x)^2 - 6 \times 3^x + 8 = 0$$

Let $y = 3^x$

$$y^2 - 6y + 8 = 0$$
$$(y - 4)(y - 2) = 0$$

Hence $y = 4$ or $y = 2$

> A substitution $y = 3^x$ is used here to obtain a quadratic equation in y. This will make the factorisation easier.

> You need to recognise that this equation has the format of a quadratic equation. Notice that 9^x can be written as $(3^2)^x$ which can then be written as 3^{2x} and $(3^x)^2$.

When $y = 4$, $4 = 3^x$

Taking logs of both sides:

$$\log 4 = \log 3^x$$
$$\log 4 = x \log 3$$
$$x = \frac{\log 4}{\log 3}$$

So $x = 1.26 \ (2 \text{ d.p.})$

When $y = 2$, $2 = 3^x$

Taking logs of both sides:

$$\log 2 = \log 3^x$$
$$\log 2 = x \log 3$$
$$x = \frac{\log 2}{\log 3}$$

So $x = 0.63 \ (2 \text{ d.p.})$

$x = 1.26$ or $x = 0.63 \ (2 \text{ d.p.})$

9 $\log_a (6x^2 + 11) - \log_a x = 2 \log_a 5$

$$\log_a \left(\frac{6x^2 + 11}{x} \right) = \log_a 5^2$$
$$\frac{6x^2 + 11}{x} = 25$$
$$6x^2 + 11 = 25x$$
$$6x^2 - 25x + 11 = 0$$
$$(3x - 11)(2x - 1) = 0$$

Hence $x = \frac{11}{3}$ or $x = \frac{1}{2}$

> Form a quadratic equation by getting all the terms onto the same side so they equal zero.

10 (a) See proof on page 121

(b) $6^{2y-1} = 4$

$$\log 6^{2y-1} = \log 4$$
$$(2y - 1) \log 6 = \log 4$$
$$2y - 1 = \frac{\log 4}{\log 6}$$
$$2y - 1 = 0.7737$$

Solving gives $y = 0.887$ to 3 d.p.

> Working should always be carried out to at least one place more than the accuracy requested; e.g. if the question asks for the answer correct to 3 decimal places (as here), then working should be shown to at least 4 d.p. before rounding to 3 d.p. at the end.

> Here you are using the fact that $\log_a x^k = k \log_a x$

BOOST

Grade ⇧⇧⇧⇧

> Answers only with no working will earn 0 marks.

> The two equations $x = a^n$ and $\log_a x = n$ given in the formula booklet have the same meaning and you need to be able to convert between the logarithm and power representations readily.

Test yourself answers

(c) $\log_a 4 = \frac{1}{2}$

$$a^{\frac{1}{2}} = 4$$
$$a = 16$$

11 (a) The initial population of the island, i.e. $N = A$ when $t = 0$

(b) The two pairs of values are substituted into the equation in turn to give a pair of equations that can be solved simultaneously.

$$100 = Ae^{2k} \qquad (1)$$
$$160 = Ae^{12k} \qquad (2)$$

Dividing equation (2) by (1) we obtain

$$1.6 = e^{10k}$$

Taking ln of both sides we obtain

$$\ln 1.6 = 10k$$

$$k = \frac{1}{10} \ln 1.6$$

$$= 0.047 \text{ (3 d.p.)}$$

(c) Substituting the value of k into equation (1)

$$100 = Ae^{2 \times 0.047}$$
$$100 = A \times 1.09856$$
$$A = 91.0283$$

When $t = 20$, $N = Ae^{kt}$

$$N = 91.0283 \times e^{0.047 \times 20} = 233$$

12 (a) See page 120 for solution.

(b) $2^{3-5x} = 12$

Taking logs of both sides

$$\log 2^{3-5x} = \log 12$$
$$(3 - 5x)\log 2 = \log 12$$
$$3 - 5x = \frac{\log 12}{\log 2}$$
$$3 - 5x = 3.5850$$
$$x = -0.117 \text{ (3 d.p.)}$$

(c) $\log_9 (3x - 1) + \log_9 (x + 4) - 2\log_9 (x + 1) = \frac{1}{2}$

$$\log_9 \frac{(3x - 1)(x + 4)}{(x + 1)^2} = \frac{1}{2}$$

$$\frac{(3x - 1)(x + 4)}{(x + 1)^2} = 9^{\frac{1}{2}}$$

$$\frac{(3x - 1)(x + 4)}{(x + 1)^2} = 3$$

$$(3x - 1)(x + 4) = 3(x + 1)^2$$
$$3x^2 + 12x - x - 4 = 3x^2 + 6x + 3$$
$$6x + 3 = 11x - 4$$
$$x = 1.4$$

Topic 7

1 (a) $y = 4x^2 + 2x - 1$

Increasing x by δx and y by δy gives
$$y + \delta y = 4(x + \delta x)^2 + 2(x + \delta x) - 1$$
$$y + \delta y = 4(x^2 + 2x\delta x + [\delta x]^2) + 2x + 2\delta x - 1$$
$$y + \delta y = 4x^2 + 8x\delta x + 4(\delta x)^2 + 2x + 2\delta x - 1$$

But $y = 4x^2 + 2x - 1$

Subtracting these equations gives
$$\delta y = 8x\delta x + 4(\delta x)^2 + 2\delta x$$

Dividing both sides by δx
$$\frac{\delta y}{\delta x} = 8x + 4(\delta x) + 2$$

Letting $\delta x \to 0$
$$\frac{dy}{dx} = \underset{\delta x \to 0}{\text{limit}}\ \frac{\delta y}{\delta x} = 8x + 2$$

(b) $y = \dfrac{8}{x^2} + 5\sqrt{x} + 1$

Putting this into index form gives:
$$y = 8x^{-2} + 5x^{\frac{1}{2}} + 1$$

Differentiating gives $\quad \dfrac{dy}{dx} = -16x^{-3} + \dfrac{5}{2}x^{-\frac{1}{2}}$

This can be written as $\dfrac{dy}{dx} = -\dfrac{16}{x^3} + \dfrac{5}{2\sqrt{x}}$

Substituting $x = 1$ gives $\dfrac{dy}{dx} = -\dfrac{16}{1^3} + \dfrac{5}{2\sqrt{1}} = -13.5$

Gradient of curve when $x = 1$ is -13.5

2 (a) $y = 4\sqrt{x} + \dfrac{32}{x} - 3$

$$y = 4x^{\frac{1}{2}} + 32x^{-1} - 3$$

$$\frac{dy}{dx} = 2x^{-\frac{1}{2}} - 32x^{-2}$$

$$\frac{dy}{dx} = \frac{2}{\sqrt{x}} - \frac{32}{x^2}$$

When $x = 4$, $\dfrac{dy}{dx} = \dfrac{2}{\sqrt{4}} - \dfrac{32}{4^2} = -1$ or -3

(b) Gradient of the tangent at $x = 4$ is -1.

Now $m_1 m_2 = -1$

Gradient of normal $= m$ so
$$m(-1) = -1, \text{ hence } m = 1$$

To find the y-coordinate of the point on the curve where $x = 4$ we substitute $x = 4$ into the equation of the curve.

$$y = 4\sqrt{4} + \frac{32}{4} - 3 = 8 + 8 - 3 = 13$$

Equation of normal having gradient, $m = 1$ and passing through the point $(4, 13)$ is given by:
$$y - 13 = 1(x - 4)$$

So $\qquad y = x + 9$

> Remember:
> $$\sqrt{4} = \pm 2$$
> which means there are two values for $\dfrac{dy}{dx}$.

> It is easier to revert back to reciprocals and roots so that the numbers can be substituted into the derivative. This enables the numerical value of the gradient to be found.

3 $y = \frac{2}{3}x^3 + \frac{1}{2}x^2 - 6x$

$$\frac{dy}{dx} = 2x^2 + x - 6 = (2x - 3)(x + 2)$$

At the stationary points $\frac{dy}{dx} = 0$

$$(2x - 3)(x + 2) = 0$$

Solving gives $\qquad x = \frac{3}{2}$ or -2

Substituting $x = \frac{3}{2}$ into the equation of the curve to find the y-coordinate gives:

$$y = \frac{2}{3}\left(\frac{3}{2}\right)^3 + \frac{1}{2}\left(\frac{3}{2}\right)^2 - 6\left(\frac{3}{2}\right) = \frac{9}{4} + \frac{9}{8} - 9 = -5\frac{5}{8}$$

Substituting $x = -2$ into the equation of the curve to find the y-coordinate gives:

$$y = \frac{2}{3}(-2)^3 + \frac{1}{2}(-2)^2 - 6(-2) = -\frac{16}{3} + 2 + 12 = 8\frac{2}{3}$$

Finding the second derivative:

$$\frac{d^2y}{dx^2} = 4x + 1$$

When $x = \frac{3}{2}$, $\qquad \frac{d^2y}{dx^2} = 7$

The positive value shows that $\left(\frac{2}{3}, -5\frac{5}{8}\right)$ is a minimum point.

When $x = -2$, $\qquad \frac{d^2y}{dx^2} = -7$

The negative value shows that $\left(-2, 8\frac{2}{3}\right)$ is a maximum point.

4 $f(x) = \sqrt[3]{x} + 2x + 5$

Writing the function in index form gives

$$f(x) = x^{\frac{1}{3}} + 2x + 5$$

Differentiating the function we obtain

$$f'(x) = \frac{1}{3}x^{-\frac{2}{3}} + 2$$

$$f'(x) = \frac{1}{3\sqrt[3]{x^2}} + 2$$

When $x = 1$, $f'(x) = 2\frac{1}{3}$

This is a positive gradient so $f(x)$ is an increasing function at $x = 1$.

Turn any negative and fractional indices back to reciprocals and roots, etc., before substituting numbers in for x.

5 (a)

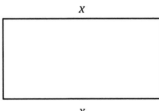

If length $= x$, then width $= \dfrac{100 - 2x}{2}$

$$= 50 - x$$

Area of pen, $A = x(50 - x) = 50x - x^2$

Differentiating the expression for this area gives:

$$\frac{dA}{dx} = 50 - 2x$$

Maximum value occurs when $\frac{dA}{dx} = 50 - 2x$

Hence $50 - 2x = 0$, so $x = 25$ m

So length = 25 m, width = $50 - x$

$$= 50 - 25 = 25 \text{ m}$$

There is only one value of x so it must be the maximum value.

We can check that it is the maximum by finding the second derivative.

$$\frac{d^2A}{dx^2} = -2$$

(The negative value above proves that the only value for x is a maximum.)

So, the length and width are the same at 25 m and the pen will need to be square.

(b) Maximum area = $x^2 = 25^2 = 625$ m^2

6 Increasing x by a small amount δx will result in y increasing by a small amount δy.

Putting $x + \delta x$ and $y + \delta y$ into the equation we have:

$$y + \delta y = (x + \delta x)^3$$
$$y + \delta y = (x + \delta x)(x + \delta x)^2$$
$$y + \delta y = (x + \delta x)(x^2 + 2x\delta x + [\delta x]^2)$$
$$y + \delta y = x^3 + 2x^2\delta x + x[\delta x]^2 + x^2\delta x + 2x[\delta x]^2 + [\delta x]^3$$
$$y + \delta y = x^3 + 3x^2\delta x + 3x[\delta x]^2 + [\delta x]^3$$

Subtracting y from both sides of the equation but on the right-hand side we subtract x^3 as this is the same as the value of y.

This gives:

$$\delta y = 3x^2\delta x + 3x[\delta x]^2 + [\delta x]^3$$

Dividing both sides by δx and letting $\delta x \to 0$

$$\frac{dy}{dx} = \lim_{\delta x \to 0} \frac{\delta y}{\delta x} = 3x^2$$

7 Increasing x by a small amount δx will result in increasing y by a small amount δy.

Substituting $x + \delta x$ and $y + \delta y$ into the equation we have:

$$y + \delta y = 2(x + \delta x)^2 - 7(x + \delta x) + 5$$
$$y + \delta y = 2(x^2 + 2x\delta x + [\delta x]^2) - 7x - 7\delta x + 5$$
$$y + \delta y = 2x^2 + 4x\delta x + 2(\delta x)^2 - 7x - 7\delta x + 5$$

But $\quad y = 2x^2 - 7x + 5$

Subtracting these equations gives

$$\delta y = 4x\delta x + 2(\delta x)^2 - 7\delta x$$

Dividing both sides by δx

$$\frac{\delta y}{\delta x} = 4x + 2\delta x - 7$$

Letting $\delta x \to 0$

$$\frac{dy}{dx} = \lim_{\delta x \to 0} \frac{\delta y}{\delta x} = 4x - 7$$

(8) Writing in index form gives

$$6x^{\frac{2}{3}} - 3x^{-3}$$

Differentiating gives

$$\left(\frac{2}{3}\right)6x^{-\frac{1}{3}} - (-3)3x^{-4} = 4x^{-\frac{1}{3}} + 9x^{-4}$$

(9) (a) $y = x^2 - 8x + 6$

Differentiating the equation of the curve to find the gradient gives

$$\frac{dy}{dx} = 2x - 8$$

At A (1, 2) the gradient is found by substituting $x = 1$ into the expression for $\frac{dy}{dx}$

Hence, the gradient of the tangent at A,

$$\frac{dy}{dx} = 2(1) - 8 = -6$$

The equation of the tangent is found using the formula:

$$y - y_1 = m(x - x_1),$$

where $m = -6$ and $(x_1, y_1) = (1, 2)$

So $y - 2 = -6(x - 1)$

$y - 2 = -6x + 6$

$6x + y - 8 = 0$

(b) The tangent and normal are perpendicular to each other, so using $m_1 m_2 = -1$ we have:

$$-6m = -1 \quad \text{giving } m = \frac{1}{6}$$

Equation of the normal with gradient $\frac{1}{6}$ and passing through A (1, 2) is:

$$y - 2 = \frac{1}{6}(x - 1)$$

$$6y - 12 = x - 1$$

$$x - 6y + 11 = 0$$

(10) (a) $y = x^3 - 3x^2 + 3x + 5$

$$\frac{dy}{dx} = 3x^2 - 6x + 3 = 3(x^2 - 2x + 1)$$

$$= 3(x - 1)(x - 1)$$

$$= 3(x - 1)^2$$

At the stationary points $\frac{dy}{dx} = 0$

$$3(x - 1)^2 = 0$$

Solving gives $x = 1$ so there is only one stationary point.

To determine the y-coordinate of the stationary point, we substitute $x = 1$ into the equation of the curve.

$y = 1^3 - 3(1)^2 + 3(1) + 5 = 6$

So, the stationary point of curve C is at (1, 6)

(b) Looking at the gradient either side of the stationary point at $x = 1$

When $x = 2$, $\frac{dy}{dx} = 3(2)^2 - 6(2) + 3 = 3$

When $x = 0$, $\frac{dy}{dx} = 3(0)^2 - 6(0) + 3 = 3$

The gradient does not change sign either side of the stationary point thus proving that the stationary point is a point of inflection.

11 (a) $y = 3x^2 - 7x - 5$

Increasing x by a small amount δx will result in y increasing by a small amount δy.

Substituting $x + \delta x$ and $y + \delta y$ into the equation we have:

$y + \delta y = 3(x + \delta x)^2 - 7(x + \delta x) - 5$

$y + \delta y = 3(x^2 + 2x\delta x + [\delta x]^2) - 7x - 7\delta x - 5$

$y + \delta y = 3x^2 + 6x\delta x + 3(\delta x)^2 - 7x - 7\delta x - 5$

But $y = 3x^2 - 7x - 5$

Subtracting these equations gives:

$\delta y = 6x\delta x + 3(\delta x)^2 - 7\delta x$

Dividing both sides by δx

$\dfrac{\delta y}{\delta x} = 6x + 3\delta x - 7$

Letting $\delta x \to 0$

$\dfrac{dy}{dx} = \underset{\delta x \to 0}{\text{limit}} \dfrac{\delta y}{\delta x} = 6x - 7$

(b) $y = ax^{\frac{5}{2}}$

$\dfrac{dy}{dx} = \dfrac{5}{2}ax^{\frac{3}{2}}$

> Remember: to differentiate you multiply by the index and then reduce the index by 1.

$\dfrac{dy}{dx} = \dfrac{5}{2}a\sqrt{x^3}$

Now $\dfrac{dy}{dx} = -2$ when $x = 4$

Hence $-2 = \dfrac{5}{2}a\sqrt{64}$

Solving for a gives $a = -\dfrac{1}{10}$

> The denominator (i.e. bottom part) of a fractional power means a root (the 2 here means a square root). The numerator in the fractional power means the power to which the number is raised. If you are unsure about indices, then look back at Topic 2.

Topic 8

1 $\displaystyle\int\left(4x^{\frac{1}{3}} - \dfrac{2}{\sqrt[3]{x}}\right)dx = \int\left(4x^{\frac{1}{3}} - 2x^{-\frac{1}{3}}\right)dx$

$= \dfrac{4x^{\frac{4}{3}}}{\frac{4}{3}} - \dfrac{2x^{\frac{2}{3}}}{\frac{2}{3}} + c$

$= 3x^{\frac{4}{3}} - 3x^{\frac{2}{3}} + c$

2 $\displaystyle\int\left(\sqrt[3]{x} - \dfrac{1}{x^4}\right)dx = \int\left(x^{\frac{1}{3}} - x^{-4}\right)dx$

$= \dfrac{x^{\frac{4}{3}}}{\frac{4}{3}} - \dfrac{x^{-3}}{-3} + c$

$= \dfrac{3}{4}x^{\frac{4}{3}} + \dfrac{x^{-3}}{3} + c$

3 $\displaystyle\int\left(\dfrac{4}{x^3} - 6x^{\frac{1}{5}}\right)dx = \int\left(4x^{-3} - 6x^{\frac{1}{5}}\right)dx$

$= \dfrac{4x^{-2}}{-2} - \dfrac{6x^{\frac{6}{5}}}{\frac{6}{5}} + c = -2x^{-2} - 5x^{\frac{6}{5}} + c$

Test yourself answers

④ $\int\left(\dfrac{2}{\sqrt{x}} - x^{\frac{3}{2}}\right)dx = \int\left(2x^{-\frac{1}{2}} - x^{\frac{3}{2}}\right)dx$

$$= \dfrac{2x^{\frac{1}{2}}}{\frac{1}{2}} - \dfrac{x^{\frac{5}{2}}}{\frac{5}{2}} + c = 4x^{\frac{1}{2}} - \dfrac{2}{5}x^{\frac{5}{2}} + c$$

⑤ $\int_0^4\left(x^{-\frac{1}{2}} + 2x\right)dx = \left[2x^{\frac{1}{2}} + x^2\right]_0^4 = \left[2\sqrt{x} + x^2\right]_0^4$

$$= \left[(4 + 16) - (0)\right] = 20$$

⑥ (a) $\int\left(\dfrac{5}{x^3} - 3x^{\frac{1}{4}}\right)dx = \int\left(5x^{-3} - 3x^{\frac{1}{4}}\right)dx$

$$= \dfrac{5x^{-2}}{-2} - \dfrac{3x^{\frac{5}{4}}}{\frac{5}{4}} + c$$

As this is an indefinite integral the constant of integration, c must be included in the answer.

$$= -\dfrac{5}{2}x^{-2} - \dfrac{12}{5}x^{\frac{5}{4}} + c$$

(b) (i) Solving the equations of the curve and the straight line simultaneously by equating the y-values gives:

$6 + 4x - x^2 = x + 2$

$0 = x^2 - 3x - 4$

$0 = (x + 1)(x - 4)$

Giving $x = -1$ or $x = 4$

From the question, point A is in the first quadrant.

Hence $x = 4$

When $x = 4$, $y = 4 + 2 = 6$

So A is the point $(4, 6)$

Don't forget that you need both coordinates for point A.

(ii) Area under the curve between $x = 0$ and $x = 4$ $= \int_0^4\left(6 + 4x - x^2\right)dx$

$$= \left[6x + 2x^2 - \dfrac{x^3}{3}\right]_0^4$$

$$= \left(24 + 32 - \dfrac{64}{3}\right) - \left(0\right)$$

$$= 34\dfrac{2}{3} \text{ square units}$$

Area under the line between $x = 0$ and $x = 4$ $= \dfrac{1}{2}(4)(2 + 6)$

$$= 16 \text{ square units}$$

Shaded area $= 34\dfrac{2}{3} - 16 = 18\dfrac{2}{3}$ square units

⑦ (a) $\int\left(\dfrac{3}{x^2} - 2\sqrt{x}\right)dx = \int\left(3x^{-2} - 2x^{\frac{1}{2}}\right)dx$

$$= \dfrac{3x^{-1}}{-1} - \dfrac{2x^{\frac{3}{2}}}{\frac{3}{2}} + c$$

$$= -3x^{-1} - \dfrac{4}{3}x^{\frac{3}{2}} + c$$

(b) (i) For the coordinates of A and B substituting $y = 0$ gives

$0 = 5x - 4 - x^2$

$x^2 - 5x + 4 = 0$

$(x - 4)(x - 1) = 0$

$x = 1$ or $x = 4$

At point A, $x = 1$ and at point B, $x = 4$

(ii) Area under curve between A and B $= \int_0^4 \left(5x - 4 - x^2\right)dx$

$= \left[\frac{5x^2}{2} - 4x - \frac{x^3}{3}\right]_1^4$

$= \left[\left(\frac{5 \times 16}{2} - 16 - \frac{64}{3}\right) - \left(\frac{5 \times 1}{2} - 4 - \frac{1}{3}\right)\right]$

$= \left[\left(2\frac{2}{3}\right) + \left(1\frac{5}{6}\right)\right]$

$= 4\frac{1}{2}$ square units

Area above curve, between B and C $= \int_4^5 \left(5x - 4 - x^2\right)dx$

$= \left[\frac{5x^2}{2} - 4x - \frac{x^3}{3}\right]_4^5$

$= -\left[\left(\frac{5 \times 25}{2} - 20 - \frac{125}{3}\right) - \left(\frac{5 \times 16}{2} - 16 - \frac{64}{3}\right)\right]$

$= 62\frac{1}{2} + 20 + \frac{125}{3} + 40 - 16 - \frac{64}{3}$

$= \frac{11}{6}$ square units

Total area $= 4\frac{1}{2} + \frac{11}{6} = \frac{19}{3} = 6\frac{1}{3}$ square units

> Using integration to find the area between a curve and the x-axis will give a positive value for regions above the x-axis and a negative value for regions below the x-axis.
>
> In order find the total shaded area, positive areas need to be added together. To ensure that areas are positive, the value of definite integrals will need to be negated for regions below the axis.

8 Solving the equations of the curve and straight line simultaneously to find the coordinates of the points of intersection A and B.

Equating the y-values gives:
$$9 - x^2 = x + 3$$
$$x^2 + x - 6 = 0$$
$$(x + 3)(x - 2) = 0$$

Solving gives $x = -3$ or $x = 2$

Substitute both values of x into the equation of the straight line to find the corresponding y-coordinates.

When $x = -3$, $y = (-3) + 3 = 0$

When $x = 2$, $y = 2 + 3 = 5$

By looking at the graph, A is $(-3, 0)$ and B is $(2, 5)$.

Area under the curve between $x = -3$ and $x = 2$ is given by:

$$\int_{-3}^{2} \left(9 - x^2\right)dx$$

$= \left[\left(9x - \frac{x^3}{3}\right)\right]_{-3}^{2} = \left[\left(9(2) - \frac{(2)^3}{3}\right) - \left(9(-3) - \frac{(-3)^3}{3}\right)\right]$

$= \left[\left(18 - \frac{8}{3}\right) - (-27 + 9)\right] = 33\frac{1}{3}$

Area of the triangle under line $y = x + 3$ is:
$$\frac{1}{2} \times 5 \times 5 = 12\frac{1}{2}$$

Hence the shaded area $= 33\frac{1}{3} - 12\frac{1}{2} = 20\frac{5}{6}$ square units

Topic 9

① (a) $3\mathbf{u} - 2\mathbf{v} = 3(3\mathbf{i} + 4\mathbf{j}) - 2(-2\mathbf{i} + 3\mathbf{j}) = 13\mathbf{i} + 6\mathbf{j}$

(b) $\overrightarrow{UV} = -2\mathbf{i} + 3\mathbf{j} - (3\mathbf{i} + 4\mathbf{j}) = -5\mathbf{i} - \mathbf{j}$

Length of line UV $= \sqrt{(-5)^2 + (-1)^2} = \sqrt{26}$

② (a) $\overrightarrow{AB} = \overrightarrow{AO} + \overrightarrow{OB}$
$$= -(4\mathbf{i} - 3\mathbf{j}) + 6\mathbf{i} + \mathbf{j}$$
$$= 2\mathbf{i} + 4\mathbf{j}$$

(b) $|\overrightarrow{AB}| = \sqrt{2^2 + 4^2} = \sqrt{20} = 2\sqrt{5}$

③ First draw a picture of a square and label the corners like this.

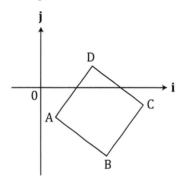

Don't worry too much about the exact positions of the corners of the square.

$\overrightarrow{AB} = \overrightarrow{AO} + \overrightarrow{OB} = -3\mathbf{i} + 2\mathbf{j} + 6\mathbf{i} - 4\mathbf{j} = 3\mathbf{i} - 2\mathbf{j}$

As $\overrightarrow{DC}$ is parallel to, and the same length as, $\overrightarrow{AB}$ the two vectors will be identical.

Hence $\overrightarrow{DC} = \overrightarrow{AB} = 3\mathbf{i} - 2\mathbf{j}$

Now $\overrightarrow{DC} = \overrightarrow{DO} + \overrightarrow{OC}$

$3\mathbf{i} - 2\mathbf{j} = \overrightarrow{DO} + 8\mathbf{i} - \mathbf{j}$

$\overrightarrow{DO} = -5\mathbf{i} - \mathbf{j}$

$\overrightarrow{OD} = 5\mathbf{i} + \mathbf{j}$

Hence position vector of D is $5\mathbf{i} + \mathbf{j}$

④ (a) The position vectors of points A, B and C are:

$\overrightarrow{OA} = -2\mathbf{i} + 5\mathbf{j}$

$\overrightarrow{OB} = 4\mathbf{i} + 3\mathbf{j}$

$\overrightarrow{OC} = 10\mathbf{i} + \mathbf{j}$

A rough sketch of the positions can be drawn.

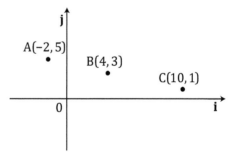

$\overrightarrow{AB} = \overrightarrow{AO} + \overrightarrow{OB} = -(-2\mathbf{i} + 5\mathbf{j}) + 4\mathbf{i} + 3\mathbf{j} = 6\mathbf{i} - 2\mathbf{j}$

$\overrightarrow{AC} = \overrightarrow{AO} + \overrightarrow{OC} = -(-2\mathbf{i} + 5\mathbf{j}) + 10\mathbf{i} + \mathbf{j} = 12\mathbf{i} - 4\mathbf{j}$

(b) $\overrightarrow{AB} = 6\mathbf{i} - 2\mathbf{j}$ and $\overrightarrow{AC} = 2(6\mathbf{i} - 2\mathbf{j})$ both vectors have the same vector part (i.e. $6\mathbf{i} - 2\mathbf{j}$) and are therefore parallel. Also, they share the same common point A, so A, B and C must lie in a straight line.

(c) $AB : AC = 1 : 2$

5 (a) $\overrightarrow{OD} = 2\overrightarrow{OA} = 2\mathbf{a}$

$\overrightarrow{OC} = \tfrac{1}{2}\overrightarrow{OB} = \dfrac{\mathbf{b}}{2}$

$\overrightarrow{CD} = \overrightarrow{CO} + \overrightarrow{OD}$

$\quad = 2\mathbf{a} - \dfrac{\mathbf{b}}{2}$

(b) The point E dividing AB in the ratio $\lambda : \mu$

has position vector $\overrightarrow{OE} = \dfrac{\mu\mathbf{a} + \lambda\mathbf{b}}{\lambda + \mu}$

As E divides AB in the ratio of $1 : 2$

it has position vector $\overrightarrow{OE} = \dfrac{2\mathbf{a} + \mathbf{b}}{1 + 2} = \dfrac{2}{3}\mathbf{a} + \dfrac{1}{3}\mathbf{b}$